SpringerBriefs on Pioneers in Science and Practice

Volume 14

Series Editor

Hans Günter Brauch

HF794117

For further volumes:
http://www.springer.com/series/10970
http://www.afes-press-books.de/html/SpringerBriefs_PSP.htm

Klaus von Beyme

Klaus von Beyme

Pioneer in the Study of Political Theory and Comparative Politics

Springer

UNIVERSITÄT HEIDELBERG ZUKUNFT SEIT 1386

Klaus von Beyme
Institut für Politische Wissenschaft
Universität Heidelberg
Heidelberg
Baden-Württemberg
Germany

The cover photograph was taken from the website of IPSA with the photos of all IPSS Presidents, at: <http://legacy.ipsa.org/site/content/view/173/92/lang,en/>. This photograph was taken by Michael Schwarz of Heidelberg University who granted permission.

ISSN 2194-3125 ISSN 2194-3133 (electronic)
ISBN 978-3-319-01534-7 ISBN 978-3-319-01535-4 (eBook)
DOI 10.1007/978-3-319-01535-4
Springer Cham Heidelberg New York Dordrecht London

Library of Congress Control Number: 2013944544

© The Author(s) 2014
This work is subject to copyright. All rights are reserved by the Publisher, whether the whole or part of the material is concerned, specifically the rights of translation, reprinting, reuse of illustrations, recitation, broadcasting, reproduction on microfilms or in any other physical way, and transmission or information storage and retrieval, electronic adaptation, computer software, or by similar or dissimilar methodology now known or hereafter developed. Exempted from this legal reservation are brief excerpts in connection with reviews or scholarly analysis or material supplied specifically for the purpose of being entered and executed on a computer system, for exclusive use by the purchaser of the work. Duplication of this publication or parts thereof is permitted only under the provisions of the Copyright Law of the Publisher's location, in its current version, and permission for use must always be obtained from Springer. Permissions for use may be obtained through RightsLink at the Copyright Clearance Center. Violations are liable to prosecution under the respective Copyright Law.
The use of general descriptive names, registered names, trademarks, service marks, etc. in this publication does not imply, even in the absence of a specific statement, that such names are exempt from the relevant protective laws and regulations and therefore free for general use.
While the advice and information in this book are believed to be true and accurate at the date of publication, neither the authors nor the editors nor the publisher can accept any legal responsibility for any errors or omissions that may be made. The publisher makes no warranty, express or implied, with respect to the material contained herein.

Printed on acid-free paper

Springer is part of Springer Science+Business Media (www.springer.com)

Preface

It is an extraordinary pleasure to present Klaus von Beyme as a *Pioneer in the Study of Political Theory and Comparative Politics* whom I have known as a scholar for over 30 years, and whom I have never ceased to like personally and to respect professionally. Let me assure you this happens rarely.

Klaus von Beyme and I struck up an acquaintance 11 IPSA congresses ago, at the 1979 World Congress in Moscow, an exercise in 'peaceful coexistence' held at Lomonosov University. When we stepped out of an elevator in the (since demolished) Rossija Hotel, Beyme's tightly packed travel bag burst open and spilled its contents on the hotel lobby's floor. Without further ado, we both knelt and started dumping his clothing back into the bag. I like to picture the present moment, which I treasure, as a sort of late sequel to that act of spontaneous collaboration.

Klaus von Beyme once wrote that, in analyzing politics between, as it happens, Madrid and Moscow, he had let himself be guided by the principle to treat each country with as much empathy as his own. No wonder: Beyme is fluent in seven, and his works have been translated into 10 languages, including Chinese, Korean, Polish, Croatian, Slowenian, Italian, Greek, and Spanish—English as a matter of course. He has written on political theories and political systems (those of the United States, Soviet Russia, Spain, Italy, and Germany—the latter work, several times updated, has meanwhile seen 10 printings), on Central-East Europe's transition from Communism, on interest groups, political parties and comparative politics, on policy fields such as health, traffic, and residential construction. And he has devoted an increasing amount of sophisticated thinking to ways in which political science relates to the social and cultural

---

This text represents an address by Rainer Eisfeld that was delivered at the Mattei Dogan Award Session on July 12, 2012, XXII World Congress of Political Science in Madrid. The permission to use this speech as a preface to this volume was granted by Rainer Eisfeld. See also the tribute at the IPSA Website at: http://archive.org/details/AwardSession-2012PrizeOfTheFoundationMatteiDoganAwardedByThe?start=3113.5

world around it, with a focus on architecture and on art. Whoever wishes to extend inquiries beyond the discipline's traditional limits, may draw encouragement from him.

Beyme was the first West German exchange student to study in Moscow during the late 1950s, and he distinguished himself as a Research Fellow at Harvard University's Russian Research Center immediately afterward. Lomonosov University made him a Honorary Professor only 2 years ago for his significant contributions both to the development of political science and to relations between that Moscow University and its German counterparts. For 7 years, Beyme served on the Research Council of the European University Institute in Florence. He was a Fellow both at the Wissenschaftskolleg Berlin and at the Maison des Sciences de l'Homme in Paris. Bearer of an honorary doctorate from Berne University in Switzerland, a former president of the German Political Science Association in the 1970s and of IPSA in the early 1980s, Beyme is a rare example of the political scientist as global scholar and public intellectual.

A public intellectual has been defined as one who seeks to advance both knowledge and human freedom. Beyme has always been extremely reticent about publicly avowing his fundamentally humanist orientation. But he did, now and then, refer to the impressions deeply imprinted on his memory of fleeing, as a 10-year-old boy, from the burning city of Breslau, only to find the city of Halberstadt, upon his arrival, equally in flames. Small wonder that he concluded a 1987 work on post-World War II architecture and urban development policies in the two German states with a remarkably unequivocal sentence. I quote: "The surviving Germans' sense of having escaped, in the Second World War, by the skin of their teeth needs to be transformed into the awareness that, in a Third World War, peoples would lose more than their cities' visual identity."

Even before he became IPSA President, Beyme supported admitting the German Democratic Republic to our organization. At the time, the East German delegates were still working miracles when traveling to IPSA conferences: They boarded their plane as jurists, economists or philosophers, exiting it as political scientists. "Change through closer ties" was Beyme's often affirmed policy. He and Secretary General John Trent even found a face-saving formula that allowed the Republic of China—at least for half a decade—to join IPSA without alienating Taiwan.

My German colleague Wolfgang Merkel has remarked about Beyme that "his theoretical creativity has always been constrained by the scruples of his enormous historical and empirical knowledge". Beyme's 1994 work on *System Transformation in Eastern Europe*, subsequently translated into English and Korean—another divided nation there—, provides a perfect example of a work saturated with conceptional and historical insight, replete with sophisticated observations. Every such study would have looked at institution building, dealt with social and ethnic cleavages and their effects on the establishment of political parties, focused on the formidable problem of synchronizing political and economic transformations. Beyme went beyond.

He provided an incisive analysis of 'transformation without elite exchange': in administration, in the economy, in academe (excepting the former GDR), but also in politics, particularly in Albania, Bulgaria, Romania, Serbia, and Russia. He argued that the concept of civil society, developed as a counter ideology to actually existing socialism, mirrored not a few intellectuals' anti-political stance and their lack of familiarity with economic matters. In Beyme's opinion, the notion's idealistic features fit the often harsh new political and economic realities badly. He identified nationalism as a much more potent force when it came to providing political legitimation and psychological gratifications under conditions of economic downturn, and he correctly foresaw that it was nationalism which, for a considerable period, would come to fill the ideological vacuum. Beyme's account of ethnic policies under socialism—"culminating in the right to establish folk dance groups", as he ironically notes—their mobilizing post-1989 consequences, and a few beginnings of new minority policies—that account is among his book's most perceptive chapters.

In another masterful, tightly argued chapter, linking the three dimensions of the political—polity, politics, and policies—Beyme explored the complex processes which had steadily eroded the legitimacy of the socialist system. However, as Beyme has also emphasized time and again: The discipline had absolutely failed to predict the collapse of existing socialism, and theories of totalitarianism, with their emphasis on terror and coercion, had impeded rather than helped along any attempts at assessing future developments. Different paradigms had not prevented their advocates either from making false assumptions. Here Beyme was characteristically candid; he himself had preferred the interest group approach pioneered by Gordon Skilling. Ever the skeptical realist, Beyme maintained then, and he remains convinced now:

Political science cannot predict processes on the macrolevel. Neither the student rebellion, nor the oil crisis, nor finally the rise of fundamentalism were forecast by the discipline. "Informed guesswork", according to Beyme, is the best we may expect.

Comparative Political Science his almost latest book published 3 years ago, assembles 21 articles and chapters from the past decade in three sections—'Comparing Theories', 'Comparing Institutions', and 'Comparing Policies'. These pieces attest to Beyme's undiminished intellectual curiosity and creativity. As an example, a single chapter must suffice here: A tightly structured review of five decades of German health policy—of the visions and the conundrums, the decisions and the nondecisions, the attempts at regulation and the barriers against such efforts, the interplay of historical inheritances, institutional structures, and organized interests. These mere 10 pages impress the reader as nothing short of brilliant.

Of the volume's chapters, three focus on cultural and art policy. When Klaus von Beyme began writing about culture and politics in 1987, he started out with a book on the part played by architecture and urban planning in the process of rebuilding the two Germanys after 1945—because he held that "no field of art is as strongly impregnated politically as architecture and urban development". More such works followed: on German cultural policy; on *The Art of Power and the Countervailing Power of Art*; on *Age of Avantgardes: Art and Society, 1905–1955*; finally on *Fascination of the Exotic: Exoticism, Racism and Sexism in Art*.

The Art of Power and the Countervailing Power of Art contained the gist of Beyme's considerations on the relationship between art, polity, and politics. I quote:

> Since the Renaissance, politics increased its autonomy by a symbiosis with art, which served the aesthetic legitimation of authority... In democracies with universal suffrage and parliamentary responsibility of governments, art and power abandoned that temporary symbiosis and began growing apart... Nowadays, aesthetical orchestration of politics pushes aside art as a technique for legitimating authority... To the extent that the state promotes art only marginally..., economics finds its way into art production.

In a chapter on "Architecture in the Service of Awe and Intimidation", Beyme identified early modern monumentalism as an expression of agonistic societal pluralism and subsequent twentieth century sites for mass rallies, to be filled with indoctrinated crowds, as a distinguishing mark of totalitarian dictatorships. In another chapter on "The October Revolution's Political Myths in the Arts", he argued that "mythologizing the collective" had been the revolution's most important integrative mechanism, on which Stalin had later been able to build his show trials.

In 2008, Klaus von Beyme received the Schader Foundation Award, one of the most important German awards given to social scientists, for distinguishing himself in the "dialog between the social sciences and practical life". Beyme's work provides an enduring incentive not to settle for political studies in the sense of a reductionist science focusing on the 'management' of parliamentary and party government. Rather, political scientists should sharpen their minds and open their hearts to addressing those pressing national, regional, and global challenges which transcend any self-imposed confines of our discipline. In concluding, I would like to refer to just a single instance indicating issues of the kind which a political science informed by Klaus von Beyme's example might be addressing more widely.

In his treatise on world poverty and human rights, Yale political philosopher Thomas Pogge has argued that Western political and financial institutions are deeply implicated in actively maintaining the corrupt and oppressive rulers of impoverished states, due to the interest of affluent democracies in obtaining access to natural resources and in issuing lucrative loans. Sufficient material may be found in Pogge's essays which the discipline might debate with a view to speaking out in favor of a more just and more equitable organization of political processes and institutions, both nationally and internationally.

It is my profound conviction that awarding the Mattei Dogan Prize to Klaus von Beyme for his outstanding achievements will provide a powerful boost to the kind of political science which does not shy away from incorporating historical dynamics, societal conflicts, and embedded power relations, and which supports men and women—wherever they may live—in their quest to participate more effectively, more knowledgeably, and more freely in today's political decision making.

Rainer Eisfeld

Contents

Klaus von Beyme in his office in the 1970s. *Source* This photograph was taken by the press office of Heidelberg University who granted permission to use it

Part I
Klaus von Beyme

Klaus von Beyme. *Source* This photograph was taken by Michael Schwarz who granted permission to reproduce it

Chapter 1
Walking on Two Legs

Comparative Politics in East and West

1.1 Personal Background

In 1995, fifty years after the end of World war II, a typical abstract debate was launched in Germany about whether 1945 meant 'liberation' or 'collapse' to the Germans.[1] I was only ten years old in 1945 and, at the time I certainly did not have a mature perception of what had happened. In spite of the disaster of being expelled from Silesia and the hardships of escaping from the Soviet occupation zone to the West, the atmosphere for us youngsters was rather challenging as in a more serious game of 'cops and robbers' which we used to play. We were the robbers and the allied troops were the cops. Our fathers were in Russian prisoner camps, so the youngsters had to organize food and timber for a mother with five children. Comparative Politics started for me in 1945: my mother gladly accepted coals stolen in a Russian camp, but she refused bread which I had obtained by breaking into a shelter of the American forces. Russians were considered as inferior and enemies; Americans and British soldiers were accepted as behaving on the whole correctly and offering prospects for a more civil and democratic life in Germany. My interest in politics dated from the occupation years, as my zeal to learn Russian and Polish in the late school years was no doubt motivated by hatred. But hatred can grow into love. Only indifference is a poor motive for comparing nations.

[1] This text was first published by © Klaus von Beyme: "Walking on Two Legs: Comparative Politics in East and West", in: Hans Daalder (ed.): *Comparative European Politics: The Story of a Profession*. London, Pinter 1997: 215–226. The permission to reproduce this text in this book was granted on 30 April 2013 by Ms. Claire Weatherhead, Permissions Manager, Bloomsbury Publishing Plc, London, UK.

K. von Beyme, *Klaus von Beyme*, SpringerBriefs on Pioneers in Science and Practice 14,
DOI: 10.1007/978-3-319-01535-4_1, © The Author(s) 2014

Carl Joachim Friedrich, who came from Harvard University and accepted a part-time job at the University of Heidelberg, founded political science in Heidelberg. He became my teacher and he certainly had the deepest impact on my development, with the exception of Karl Deutsch whom I first met a MIT. Since the late 1950s I have divided my interests between East and West. Political Science was almost inexistent in the winter term when Friedrich taught at Harvard. So I went to 'SciencesPo' in Paris, working mostly with Maurice Duverger and Raymon Aron. When Adenauer established diplomatic relations with Moscow he negotiated not only the return of the last German prisoners of war but also a culture exchange of students. I went with the first crew of exchange students for one year to the Lomonosov University in Moscow. There was no political science and no possibility in 1959–1960 to work on the Soviet Union. I had to choose a topic on the history of political theory in Czarist Russia (von Beyme 1965). My Russian tutor was a professor of the history of economic ideas. Political Scientists at that time were sent to the Department of Economics, no doubt a useful experience for studying the Soviet system. Not only political theory—if it existed at all—was highly influenced by political economy. The whole system was determined by economic efforts. It is not by chance that my largest comparative book on communist countries later dealt with *Economics and Politics within Socialist Systems* (von Beyme 1982). When I came back in 1960 Carl Friedrich recruited me for an academic career and organized a scholarship at the Russian Centre of Harvard University.

Friedrich wanted me to become a specialist in Soviet affairs. He even invited me to work with him on a new edition of his seminal book *Totalitarian Dictatorship and Autocracy* (Friedrich/Brzezinski 1956, 1965). Soon our different views became between us: for one, I wanted to "walk on two legs", I refused to submit a book on Soviet Federalism which I wrote for Friedrich's series on federal systems, to the faculty as a 'second doctorate' (Habilitation) required for budding scholars to become a professor (von Beyme 1964a, b). I turned instead to a comparative study on *Parliamentary Systems in Europe* (von Beyme 1970, English in: 2000). We had also different perceptions concerning totalitarianism. Abundant travels in all the Socialist countries had convinced me that there was a certain development in communist countries towards 'authoritarian rule with minimal pluralism'. Apparently Brzezinski, the co-author of the first edition of *Totalitarian Dictatorship*, also developed views different from Friedrich's. The theory of totalitarianism involved the idea that "one possibility should be excluded, except in the satellites: the likelihood of an overthrow of the regimes by revolutionary action from within" (Friedrich/Brzezinski 1956: 375). Brzezinski pulled out, and the second edition was revised by Friedrich alone. In his contribution to the 'Friedrich Festschrift' Brzezinski wrote on "Dysfunctional Totalitarianism". Without attacking his former teacher he gave voice to his dissent on many details in the evaluation of totalitarian communist systems (in: von Beyme 1971a: 375–389).

Box 1: Brief Curriculum Vitae of Klaus von Beyme

1934	Born in Saarau, Silesia
1954–1956	Apprentice in a publishing house in Brunswick
1956–1961	Studies of political science, sociology, history and history of art at the Universities of Heidelberg, Munich, Paris
1959–1960	Exchange student at Moscow University (MGU)
	Research fellow at the Russian Research Center, Harvard University
1963–1967	Assistant of Professor Carl J. Friedrich
1967–1973	Professor of political science at Tuebingen
1971	For a short time rector of the University of Tuebingen
1974–1999	Professor of political science at Heidelberg
1973–1975	President of the German Political Science Association
1979	Visiting professor at Stanford University
1982–1985	President of the International Political Science Association
1983–1990	Member of the Research Council of the European University Institute, Florence
1985	Visiting professor at the École des sciences politiques, Paris
1988	Member of the Academia Europaea
1989	Visiting professor at Melbourne University
1992–1993	Visiting professor at the Science Center, Berlin
1993–1998	Member of the scientific board at the Wissenschaftskolleg in Berlin
1995	Honorary member of the Humboldt-University in Berlin
1995 ff.	Member of the Academy Berlin-Brandenburg
2001	Honorary doctorate, University of Berne, Switzerland
2001	Fellow at the 'Wissenschaftskolleg' Berlin
2005	Fellow at the 'Maison des Sciences de l'homme', Paris
2008	Schader award for applied social sciences, Darmstadt
2010	Honorary professorship at the Lomonossov University in Moscow
2012	Mattei Dogan reward for "high achievements in Political Science" of the International Political Science Association in Madrid

See for additional biographical information on the author, his publications, vuídeos, awards also the website on this book: <http://www.afes-press-books.de/html/SpringerBriefs_PSP_vonBeyme.htm>

1.2 Studies on Western Democracies

There was no chair on Soviet politics in Germany at that time, and I was advised to turn to comparative politics in East and West. After my 'Habilitation' in 1967, I accepted a chair in comparative politics at the University of Tübingen, where I stayed—next to Theodor Eschenburg the great instructor of Germans in the high-brow press after war—from 1967 to 1974, after which I returned to Heidelberg. At Heidelberg I was not the legal successor to Carl Friedrich, but to the second holder of a chair in the department, Dolf Sternberger, another star in the German high-brow press. He had induced me to engage in parliamentary studies, but not because I agreed with him. I was rather opposed to his devoted Anglophile views. Sternberger fought all his life for a 'British System' in Germany. Under the Grand

Coalition (1966–1969) in Bonn he even launched a campaign for the introduction of the British system of plurality vote. I was, instead, rather interested in disputing the ideology which was widespread after the war in Germany that 'true parliamentarianim' of the British type should be contrasted to the French type of 'government d'assemblée'. The remnants of 'consociationalism' in Germany, developed in a long tradition of religious pluralism with West Germany (excluding Prussia) shared with the Netherlands, Switzerland and Austria, led me to become more interested in non-majoritarian systems. The project on "Small European Democracies" and contacts with Hans Daalder, Arend Lijphart, Stein Rokkan and Gerhard Lehmbruch, made the younger generation of political scientists in Germany more skeptical about the uncritical Anglomania of many scholars of the older generation. Only Carl Friedrich was free of it. He always had a strong interest in Switzerland, though he studied this country rather from the institutional point of view and hardly in the perspective of political sociology and political culture studies. In many historical studies on parliamentary systems since the French Revolution I discovered very different traditions of parliamentary government, especially in Scandinavia and the Benelux-states. A stay for several weeks in the *Institut for Statskunskap* under Nils Stjernquist in Lund was very decisive for the pluralization of my views.

The methodological problem arising in these comparative studies was that *Anglomania* in no way was limited to Germany liberals. The British model was influential via diffusion of ideas in many other countries which lacked the social conditions for imitating the British model. A by-product of this kind of question was the study of competing models, especially of the American system on European constitutionalism (von Beyme 1987c). The functionalist debate at that time made me alert to the fact that the old anthropological debate, whether institutions spread via 'diffusion' or spring up as 'functional equivalents' without direct impact from abroad, was pertinent to the development of political systems in comparative perspective. *Semi-presidential systems* and certain borrowings from the American models in Europe were frequently rather a search for the solution of a problem which created functional equivalents than a direct impact of a foreign device.

I had a parallel experience in another field of comparative politics, one which most political scientists left rather to sociologists: that of industrial relations. In Germany there was no tradition of studies on industrial relations in Political Science, though the term '*Arbeitsbeziehungen*' was budding in the political debate under the rise of the Social Democrats into a governmental party. The problem on the continent in the roaring late 1960s and early 1970s was that trade unions were mostly studied in an ideological way. After the failure of the student's rebellion the anarchical leaders of the movement turned into cadres for militant 'mini-parties'. They had no political success, but many of their leaders had some impact on the public debate and later on the 'Green Party'. During the 'hot autumns', trade unions became for a while the hope for new forms of class struggles. In the mid-1970s I turned to comparative studies on labour relations. Again, I was fascinated by the plurality of systems, which had resisted the unifying impact of Marxist ideology in many countries. The model of German 'co-determination' was criticized by the class-conscious researchers in the field. Scientific truth came rather to conclusion similar to those of a Communist French trade unionist, a leading member of the CGT, who in a panel discussion complained:

> On Sunday, we fundamentally criticize the German trade unions for their course of integration and adaptation to the system. On Monday, however, we have to go back to our factories and dream of a position which the German trade unions have in their factories: being strong and united and being accepted by the employer's side and the government.

My study on *Trade Unions and Industrial Relations* (1980) was the first attempt in Germany to evaluate the various systems and to study trade unions and their role in different systems. With the smoothing down of ideological zeal more leftist researchers in Germany, too, slowly accepted my 'industrial relations perspective' on the topic.

The study of interest groups (von Beyme 1969, 1980) and trade unions (1977, Engl. 1980) was embedded in activities of these organizations in the political systems. The missing link between interest groups and the political systems were quite naturally 'political parties' (von Beyme 1977; Engl. 1980). It was hardly possible to be original in the field of party studies, since parties were almost the only field which became exclusively the domain of political scientists. Only later constitutional lawyers have begun to deal with this topic more frequently, at least in those countries where parties are mentioned in the constitution. The German 'catch-all-parties' were evaluated since Kirchheimer and others according to the 'Urpartei SPD' in Germany: strong organization, strong ideology, strong social links, strong leaderships under 'Kaiser' Bebel who led his 'attentive revolutionaries' like the emperor led the Prussian Army.

I opposed the nostalgia for parties of class struggle. Comparisons mostly relied on the major countries. I wanted to deal with party systems in a truly comparative perspective, including the smaller countries, including the results in the various national languages, largely neglected by the American literature, and avoiding doomsday scenarios about the decline of parties as a kind of reaction to the disappointments felt towards parties when they failed to transform the whole society. A functionalist approach showed that certain functions of parties have declined, such as representation of segments of society and the mobilization of voters. Other functions, such as the recruitment function, are today more important than they used to be in the old days of the 'parties of notables' on the one hand and the leftist 'mass parties' on the other hand.

In the early 1990s the critique of the party state in many European countries led to a global rejection of parties. The links between the political élite (denounced as 'the political class') and the party state had to be studied more thoroughly ta former élite studies—including my own (von Beyme 1971, 1974) had done. I came to the conclusion that there is hardly a decline of parties going on in most countries, as conservative populists (including even our former Federal President von Weizsäcker) or Green leftists postulated. The Italian example had few consequences in the German party state, so often compared with Italy. The 'Lega ideology', which claimed that the *party nomenclatura* of the old system would wither away as the nomenclatura did under Gorbachev in the communist countries, was hardly tenable (von Beyme 1995). The later Italian decline of the old party system under Berlusconi fortunately was an exception in Europe.

My attempts to walk on two legs seemed to have little in common with findings in Western and in Eastern communist systems. But the collapse of communism showed in East and West to what extent the scientific topics in comparative politics are shaped by historical events: trade unions were 'out' as a topic in the late 1980s. Parties—for a long time considered as a topic of 'grandpa's political science' by the progressivists—were 'in' again. Party developments in East and West have been compared since 1989. In retrospective, my East European leg had to be re-evaluated (von Beyme 1995). My two areas of interest, Western democracy and communist autocracy, were not so unrelated as they may seem. Both areas were increasingly linked by a common outlook: the *policy approach.* As Chairman of the *International Political Science Association* (IPSA) (1982–1985) I tried to promote a combination of institutional studies and the new policy approach so popular in America, but hardly internalized by most European scholars in the early 1980s (von Beyme 1986a). The policy approach was particularly fruitful in communist studies because it helped to avoid endless repetitions of the institutional setting of repression in totalitarian systems and to evaluate fairly certain successes in the policy field.

The outcome orientation could be practiced even by a scholar in the field of international science policy involved in the activities of IPSA. Karl Deutsch and Candido Mendes, my predecessors, as well as my successors, Kenhide Mushakoji and Guillermo. O'Donnell, have tried a common line in dealing with the conflicts in the world. Unlike other scientific organisations, IPSA, so close to the political conflicts, was never threatened by disruption as sociological and psychological world associations, thanks to our course of mutually respecting the vital sphere of the respective powers. We have frequently been criticized for our 'appeasement policy'. In one case, that of China, it failed for a while. Without the cooperation of Georgii Shakhnazarov, who later became one of the chief advisors of Gorbachev and did a lot to 'social-democratize' the outlook of top Soviet leaders in world politics, this course might have failed more frequently. In the long run this policy and the agreement-oriented approach paid off.

1.3 Studies in Communist Systems and the Transition to Democracy

My main criticism of the Sovietologists was that they had lost touch with their discipline and the new approaches developed in comparative politics. The study was mainly limited to official data—frequently distorted ones—and the 'informed guesswork of Kremlinology'. A certain boredom attaching many Soviet Studies could not be overlooked. New approaches, as developed by Gordon Skilling in the

United States, using an interest group approach in the context of socialist countries were applied by myself at the same time. The Nestors of German Sovietology, Boris Meissner and Richard Löwenthal, attacked me vehemently for abandoning the totalitarian paradigm and even deleted certain critical remarks in the printed version of a presentation of 1966 on totalitarianism (von Beyme 1968). When I published a comparative and developmental comparison of communist countries (1975, Engl. edition 1982) the German 'Ostforschung' was mostly hostile because of my criticism of normative anti-communist credos which were the norm for dealing with the communist phenomenon. I went too far even for Peter Ludz, the most enlightened specialist on the German Democratic Republic. These minor quarrels with Ludz did not prevent the perception from outside that both of use were 'embellishing the red system' with the assumption that the communist countries were developing towards 'consultative authoritarianism'. Both of us got the unusual honor of an attack by the most conservative party leader of that time, Franz Josef Strauss in a constitutional debate in the German Bundestag. He read passages from the books of Ludz and mine and asked the Social Democrats: "These are scholars not from East Berlin or Leipzig, but from Munich and Heidelberg and belong to your party. How do you dare to ask us why we state: 'Freedom instead of socialism?'." Some years later Jürgen Habermas wanted to hire me for a triumvirate with Dahrendorf and himself in the Max-Planck-Foundation. The Foundation declined and Habermas told me: "You don't image how leftist your reputation is—you, whom I would at the utmost call a Social democratic centrist." 'Centrism' was a good classification for me, I felt always between the extremes of left and right.

Our positivist attitude towards communism started from the assumption that the system works, that some of its underlying premises are acceptable though probably in the long run not feasible, and that consensus of the population was increasingly organized via 'Goulash- or Polski-Fiat-Communism' and material benefits. Most of the Anglo-Saxon Sovietologists and some of the European specialists believed in the stability of the system. Most of us were, however, not as blind as the leading GDR specialist Gert-Joachim Glaessner, who in the month of the breakdown of the GDR published a book beginning with the introductory assumption that nobody doubted any longer that the GDR had become 'a legitimized system'. Fortunately, I could not be included in the "gallery of scholars who failed in their predictions" (Hacker 1992), because they did not take seriously the fact that the majority of the East German population was still thinking in terms of one national state. A very unscientific sentiment saved me from errors, because I was myself—as most of those Germans born east of the Oder-Neisse-line—a German patriot believing in reunification one day. My only scientific reason for this kind of national optimism was the increasing dominance of the West German media in the GDR which might stir hidden national feelings as soon as the system underwent a deep crisis. This happened in 1989.

I shared the assumption that the Soviet Union was in crisis. But like most other scholars in the field, I did not anticipate its breakdown. Those who did, offered, however, the wrong reasons: "War with China" (Amalrik) or "revolution of the nationalities" (Carrère d'Encausse). Systems change in 1989 occurred while I was away on a conference of East European specialists in Australia. The colleagues lamented: "we are ruined. We have got to apply for inclusion in the history department." Turning to me, the only one working on Eastern and Western systems they added: "You are in a good position. Not all of your books are made rubbish by history." I was not sure that my bestselling book on the *German political system* (11th edition 2010, English in 1983) would not be greatly affected by this systems change. It looked like the formation of a confederacy where new rules of the political game were negotiated by the two German states. Less than two years later, I was able, however, to issue the book under the same title with added chapters on the impact of West German institutions and policies in East Germany. Again we did not anticipate how complete the collapse of the separate East German system would be (von Beyme 1996).

Experiences of individual scholars are interesting only in the light of their consequences for the discipline of comparative politics. Looking back we have underrated many developments in the East. The more leftist writers on the Soviet Union were hopeful about the innovative capacity of socialist countries after the reforms of Khrushchev and the early Brezhnev era. Gorbachev (1989) has fixed the moment of decline in the mid-1970s. The oil crisis caused a deep crisis in the West. It overcame this astonishingly quickly, while the communist camp leaned back in self-righteous attitude and hailed 'the final crisis of Imperialism'. It overlooked that the satellites were increasingly affected by the oil crisis. The Soviet Union immediately prevented the imposition of new prices on its camp. But in the long run its resources got scarce and it had to hand over part of the prices on the world market to the COMECON-countries (von Beyme 1987c).

The perestroika since 1985 led to a strange reversal of attitudes: conservative normativists in the field, such as Boris Meissner in Germany, hailed Gorbachev and believed enthusiastically in his success because they were convinced that 'men make history'. The liberal positivists—calling themselves with Popper and Hans Albert 'critical rationalists'—like myself fell into pessimism for structural reasons and their knowledge of 'revolutions from above' that failed (von Beyme 1988b, 1989). The collapse of communism in 1989 had quite a different impact on the scientific community: the conservatives fell into a mood of opening bottles of champagne. No self-criticism afflicted them that their assumption of stability might be wrong. The more scientific positivists, on the other hand, experienced a crisis of the discipline: how was it that black Friday was possible? How was it possible that precise scientistic approaches—which emphasize a good prognosis, rather than a correct description of the situation (cf. Downs 1957: 21) proved unable to predict the collapse of the socialist camp? Even the conservative normativists shared the bias of positivism that prognosis should be avoided. The extrapolation of trends in a scientific prognosis on the basis of quantitative studies of indicators was close to this type of scholarship. Mises's belief in the 'infeasibility of socialism' even as model was never taken seriously. Only after the collapse this book was reprinted and widely read. There was only one school of social theory in the 1980s which developed an argument consistent with the *autopoetic version of systems theory,* which held socialism was no longer feasible within a predominant world market society. But the circle around H. R. Maturana and F. J. Varela was mostly not taken seriously by comparative politics at this time.

Systems change fascinated me from the outset—before it became a fashion. My first study in this area was on Spain. Parallel to developments in Western autocracies and Eastern dictatorships became an interesting topic for me very early—in cooperation with Dante Germino (von Beyme 1974b). The liberalization process in the era of Khrushchev, for me offered hope of new developments towards the more authentic participation and well-being of the citizens in East European countries. Khrushchev's attempt to liquidate Stalinism was, however, halted halfway. It is obvious that even this author overrated the possibilities for liberalization in Eastern Europe given the backlash of Neo-Stalinism in some countries. Comparisons between the *consultative authoritarianism* in the last days of the Franco Regime in Spain and some 'People's Democracies' in the East seemed to be fruitful. In Spain, many predictions of future development towards liberalization and democratization proved to be correct (1971c). There were, however, errors in some details: I was in good company with Juan Linz, the scholar who did most to develop my interest in the transition from authoritarianism to democracy in general and my work on Spain in particular in wrongly predicting that a Christian Democratic Party might be the natural leading party after systems change in Spain.

1.4 Implications of Political Theory in the Comparative Perspective

I never dared to follow my teacher Carl Friedrich on the path of becoming an all-round scholar, who combined political theory and comparative politics on equal footing. But teaching in small departments in Europe involves certainly *multi-functionalism.* Friedrich succeeded in transmitting his theoretical interests also to those of his students who did not become professors of political philosophy. I started my critical reception of postmodern thought only after the collapse of socialism (von Beyme 1991a, b, 1992: 229ff). I remained, however, rather critical of applying new paradigms from biology and physics to the breakdown of authoritarianism. The theories of fluctuations and chaos did not, so far, exceed a rather metaphorical application to the field of systems change. Moreover, useful predictions were no reason to accept wholesale the postmodern theories of the autopoetic vesion of systems theory. This new paradigm, which since Niklas Luhman had begun to dominate large parts of macro-sociological theory in Germany, as a kind of substitute for the structural assumptions of Neo-Marxism, which petered out in the 1980s, was extremely hostile towards theories of political action.

Political Science—this was my hypothesis—will remain centered in the field of *action theories*. Autopoetic systems theories do not deny that many processes are steered by political actors, of even the 'state'. But since theories start from the assumption that acting and steering are only possible within the code of various subsystems, they underrate political leadership. Processes influenced by political decision-makers are handed over to 'history', which only ex-post facto can translate influences from one subsystem (e.g. the political one) to others (such as the economic and cultural spheres). Political Science in my perception specializes in *middle-range theories.* On the level of restrictions, it will have to take into consideration certain insights of macro-theoretical systems theories. But political science will stubbornly continue to work on theories which are centered on the actor's side. Political science and comparative politics, will, in the case of theories of systems change, also consider the first insights of *theories of fluctuation and chaos*. But our discipline should not become too fashionable, by adopting metaphors from biology and physics which have not yet proved their usefulness in the context of social sciences (von Beyme 1992, 1996). There are two traditions of comparative politics deriving form Weber and Durkheim. The first tradition normally arrives in the end at historical typologies. The second is more theory-oriented, operationalizing indicators for quantitative measurement. The first method has been considered inferior in its intellectual status by quantification-minded scholars favoring correlational methods. Smelser (1973) once called the Weberian type a 'method of systematic comparative illustration'. The number of cases compared tends to be small and the possibility establishing systematic control over the sources of variation in social phenomena is reduced. Scientific generalizations seem hardly to be possible. But new approaches to comparative social research (cf. Ragin 1989: 15ff) have shown that the comparative method can be superior to the variable-oriented statistical methods in several ways: the piecemeal manner with no access to holistic notions. The traditional comparative method reflects irregularities and deviant cases. The comparative method in the Weberian tradition forces the researcher to become familiar with the cases relevant to the analysis, whereas the statistical methods only disaggregate cases into variable without entering into the analysis of individual differences among them. The normal goal of a variable-oriented investigation is to produce generalizations about relationships among variables, not to understand specific historical outcomes. I have remained interest in both.

1.5 Conclusion

Science is mortal, the more scientific a work, the quicker it seems to be dated. Even the exact natural sciences experience this every day. But political science is under a more serious threat: it is dated not only by new results but also by shifting fashions as regards the questions asked. The collapse of regimes leads to a situation in which scientific results are of only historical interest, mostly no longer able to compete with the results of historical studies. Once the archives are open, most political science assumptions and prognoses are exploded by shells of counter-evidence from sources so far unknown. Only jurisprudence has a still smaller life-expectancy as regards its findings. The proverb: "One little act of the legislator can obliterate whole libraries every day" comes to mind. Some of my more historically-oriented early books had a longer life than later more theory-oriented works. But in the case of Friedrich I was shocked by the experience that among his important books the work on 'Totalitarian Dictatorship'—which was mostly wrong in its assumptions—after his death was more frequently cited than his innovative books.

A meeting with Pope John Paul II in the Vatican in 1998. *Source* This photograph is from the author's personal photo collection

The change of fashions does not even need the collapse of a system. Colin Crouch in Germany was more successful to promote his idea of a decline of democracy towards a system which he called 'postdemocracy' than in Britain. Again I felt challenged and wrote a little book *From Postdemocracy to Neo-Democracy*. My experience in the history of art showed me that most 'post styles', from post-impressionism to 'post-dadaism' after a while were revived by 'Neo-Impressionism' and 'Neo-Dadaism'. I was not sharing the view that there ever existed an ideal phase of democracy—especially in Germany, though some writers embellished the Adenauer era as the highlight in German democracy. Crouch recognized correctly certain processes of decline, especially in the sphere of interest groups, but mostly overlooked innovative processes such as 'democracy in the media', positive forms of populism, the renewal of debates on the reform of democratic institutions and the revival of *neo-normative debates on social justice* in a system that might be called 'neo-democracy'.

My excursions into architecture and politics (on the reconstruction of German cities after 1945, 1987a), on art and politics in the age of the avant-gardes of modernity (1998, 2005) and the impact of colonialism and decolonization on the arts under the cover of 'Exoticism'(2008) were the expression of my original wish to study history of art and history. By family, however, found this daring risky and advised to study 'something useful', such as law and with less acceptance social sciences. My favorite subjects remained sub-subjects in my studies, but came to the center of my activities as soon as I entered the status of an emeritus. History for my work was increasingly the history of political ideas (2002). I tried this subject in overcoming the narrow bias of normal books in the style 'From Plato to Habermas' in favor of a comparative outlook reaching again from Russia to Spain, without neglecting German history where I saw a chance to offer the first overview on the political theories from the Middle Ages to the present time (2009).

Klaus von Beyme on 9 October 2009 at the Wissenschaftszentrum Berlin with two of his most prominent students, Prof. Dr. Andreas Busch, University of Göttingen (on the *left*), and Prof. Dr. Wolfgang Merkel, Wissenschaftszetnrum and Humboldt University of Berlin (on the *right*). The permission to reproduce this photo was granted by the photographer, Mr. Detlev Schilke (Berlin), and the costs for the reproduction were covered by the Wissenschaftszentrum Berlin (WZB)

When I started working with my teacher Friedrich I decided never to write such a general book such as *Man and his Government*. But probably my books on *Parliamentary Systems* (1970, ^{3}1999, Engl. 2000), *History of political ideas in the era of ideologies* (1789–1945) (2002), or on *Economics and Politics within Socialist Systems* (1982) are the modernized versions of such tendencies of a *generalist view*—though they contain a lot of quantitative data and tables which Friedrich tended to avoid. Mattei Dogan (and Pahre 1990) one of my closest cooperators in my time in the International Political Science Association, mentioned this author—together with scholars such as Richard Rose and Renate Mayntz—as an exception from the rule that creativity normally develops at the intersections of various disciplines. I am afraid that there are hardly exceptions to this rule. I felt most comfortable when intruding on other disciplines: history, constitutional law (especially on Constitutional Courts in Germany and Russia), economics, sociology, or even the history of art.

Klaus von Beyme as a 'public intellectual 'lecturing at the Karlsruhe Institute of Technology (KIT) in Karlsruhe (Germany) [Photo MG 9101Centre for applied Cultural Sciecne and Studium Generale, witner Semestre 2009/2010. *Source* This photo was taken from KIT's website at: <http://www.zak.kit.edu/fotos/Colloquium%20Fundamentale/WiSe%200910/Prof.%20Dr.%20 Dr.%20h.c.%20Klaus%20von%20Beyme/index.php>

Klaus von Beyme during a panel public discussion with Michael Buselmeier in late April 2008. *Source* This photo was taken from *Stadtblatt*, Amtsanzeiger of the City of Heidelberg, vol. 16, issue 18, 30 April 2008; at: <http://ww2.heidelberg.de/stadtblatt-online/index.php?artikel_id=3736&bf=>
© Stadt Heidelberg who grante permission to reproduce this photo

References

Beyme, Klaus von, 1964a: *Föderalismus in der Sowjetunion* (Heidelberg: Quelle & Meyer).

Beyme, Klaus von, 1964b: "Federal Theory and Party Reality in the Soviet Union", in: *Public Policy*, 13: 395–412.

Beyme, Klaus von, 1965: *Politische Soziologie im Zaristischen Russland* (Wiesbaden: Harrassowitz).
Beyme, Klaus von, 1968: "Gesellschaftliche Organisation und Interessenpluralismus in der Sowjetunion, in: Löwenthal, Richard; Meissner, Boris (Eds.): *Sowjetische Innenpolitik* (Stuttgart: Kohlhammer): 39–48.
Beyme, Klaus von, 1969, ^{5}1980: *Interessengruppen in der Demokratie* (Munich: Piper).
Beyme, Klaus von, ^{3}1970: *Die Parlamentarischen Regierungssysteme in Europa* (Munich: Piper) [English translation, 2000: *Parliamentary Democracy* (New York: St.Martin's Press)].
Beyme, Klaus von, 1971a: *Theory and Politics—Theorie und Politik. Festschrift zum 70. Geburtstag von Carl Joachim Friedrich* (The Hague: Nijhoff).
Beyme, Klaus von, 1971b: *Vom Faschismus zur Entwicklungsdiktatur. Machtelite und Opposition in Spanien* (Munich: Piper).
Beyme, Klaus von, 1971, ^{2}1974a: *Die politische Elite in der Bundesrepublik Deutschland* (Munich: Piper).
Beyme, Klaus von, 1974b: "Authoritarian Regimes: Developing Open Societies?", in: Germino, Dante; von Beyme, Klaus (Eds.): *The Open Society in Theory and Practice* (The Hague: Nijhoff): 109–120.
Beyme, Klaus von, 1977: Gewerkschaften und Arbeitsbeziehungen in kapitalistischen Landern. München, Piper 1977.
Beyme, Klaus von, 1980: *Challenge to Power. Trade Unions and Industrial Relations in Capitalist Countries* (London: Sage).
Beyme, Klaus von, 1982: *Economic and Politics Within Socialist Systems* (New York: Praeger) (German editions 1975, 1977).
Beyme, Klaus von, 1983: *The Political System of the Federal Republic of Germany* (New York: St. Martin's Press) (German editions: 1979–2010, 11th edition).
Beyme, Klaus von, 1985: *Political Parties in Western Democracies* (New York: St. Martin's Press) (German editions 1982, 1984).
Beyme, Klaus von, 1986a: "Plea for Policy Analysis", in: *Policy Studies Journal*, 4: 540–544.
Beyme, Klaus von, 1986b: "Attitudes of German Youth Toward Relations between the Two German States", in: *West Germany, East Germany and the German Question* (Baltimore: American Institute for Contemporary German Studies, Johns Hopkins University): 33–43.
Beyme, Klaus von, 1987a: *Der Wiederaufbau. Architektur und Städtebaupolitik in beiden deutschen Staaten* (Munich: Piper).
Beyme, Klaus von, 1987b: *The Soviet Union in World Politics* (New York: St. Martin' Press) (German editions 1983, 1985).
Beyme, Klaus von, 1987c: *America as a Model. The Impact of American Democracy in the World* (New York: St. Martin's Press) (German edition 1976).
Beyme, Klaus von, 1988: *Reformpolitik und sozialer Wandel in der Sowjetunion* (Baden-Baden: Nomos).
Beyme, Klaus von, 1988a: *Der Vergleich in der Politikwissenschaft* (Munich: Piper).
Beyme, Klaus von, 1989: "Reform and Regional Policies in the Soviet Union", in: *The Journal of Communist Studies*: 267–284.
Beyme, Klaus von, 1991a: "The Modernization of Socialism. Glasnost versus Perestroika?", in: Rustow, Dankwart, A.; Erikson, K. P. (Eds.): *Comparative Political Dynamics* (New York: Harper Collins): 262–291.
Beyme, Klaus von, 1991b: "Postmodernity, Postmaterialism and Political Theory", in: Reif, K.; Inglehart, Ronald (Eds.): *Eurobarometer. The Dynamics of European Public Opinion* (London: Macmillan): 309–333.
Beyme, Klaus von, ^{2}1992, ^{4}2007: *Theorie der Politik im 20. Jahrhundert. Von der Moderne zur Postmoderne* (Frankfurt: Suhrkamp).
Beyme, Klaus von, 1993, ^{2}1995: *Die politische Klasse im Parteienstaat* (Frankfurt: Suhrkamp).
Beyme, Klaus von, 1996: *Transition to Democracy in Eastern Europe* (London: Macmillan) (German edition 1994).

Beyme, Klaus von, 1998: *Die Kunst der Macht und die Gegenmacht der Kunst. Studien zum Spannungsverhältnis von Kunst und Politik* (Frankfurt: Suhrkamp).
Beyme, Klaus von, 2002: *Politische Theorien im Zeitalter der Ideologie. 1789–1945* (Wiesbaden: Westdeutscher).
Beyme, Klaus von, 2005: *Das Zeitalter der Avantgarden. Kunst und Gesellschaft 1905–1955* (Munich: Beck).
Beyme, Klaus von, 2008: *Die Faszination des Exotischen. Exotismus, Rassismus und Sexismus in der Kunst* (Munich: Fink).
Beyme, 2012: *Kulturpolitik in Deutschland* (Wiesbaden: VS Verlag für Sozialwissenschaften).
Beyme, Klaus von, 2013: *Von der Postdemokratie zur Neodemokratie* (Wiesbaden: Springer).
Dogan, Mattei.; Pahre, R, 1990: *Creative Marginality. Innovation and Intersections of Social Sciences* (Boulder: Westview).
Downs, Antony, 1957: *An Economic Theory of Democracy* (New York: Harper).
Friedrich, Carl. J.; Brzezinski, Zbignew, 1956, ^{2}1965: *Totalitarian Dictatorship and Autocracy* (Cambridge, Mass.: Harvard University Press).
Gorbachev, Michail, 1989: "Sotsialisticheskaya idea i revolytionnaya perestroika", in: *Pravda*, 26.
Hacker, Jens, 1992: *Deutscher Irrtümer. Schönfärber und Helfershelfer der SED-Diktatur im Westen* (Berlin: Ullstein).
Ragin, C. C., 1989: *The Comparative Method* (Berkeley: University of California Press).
Smelser, N. J., 1973: "The Methodology of Comparative Analysis", in: Warwick, D.; Osherson (Eds.): *Comparative Research Methods* (Englewood Cliffs: Prentice Hall): 101–117.

Chapter 2
The Author's Major Publications

2.1 Authored Books[1]

Der Föderalismus in der Sowjetunion. Heidelberg, Quelle & Meyer, 1964. (Short English version in: *Public Policy* 1964).

Politische Soziologie im zaristischen Russland. Wiesbaden, Harrassowitz 1965.

Das präsidentielle Regierungssystem der Vereinigten Staaten in der Lehre der Herrschafts formen. Karlsruhe, C. F. Müller 1967, enlarged edition: *Vorbild Amerika. Der Einfluß der amerikanischen Demokratie in der Welt*. München, Piper 1968. (Engl.: *America as a Model. The Impact of American Democracy in the World*. New York, St. Martins' Press 1987).

Die verfassungsgebende Gewalt des Volkes. Tübingen, Mohr 1968. (Greek.: *Syntaktikh efoysia toy laoy*. Athen, Ekdoseis 1987).

Politische Ideengeschichte—Probleme eines interdisziplinären Forschungsbereichs. Tübingen, Mohr 1969.

Interessengruppen in der Demokratie. München, Piper 1969; 5th edition 1980 (Span.: Los grupos de presión en la Democracia. Buenos Aires, Belgrano, 1986).

Die parlamentarischen Regierungssysteme in Europa. München, Piper 1970; 2nd edition 1973. 3rd edition: *Die parlamentarische Demokratie. Entstehung und Funktionsweise*. Opladen, Westdeutscher Verlag 1999 (Engl.: *Parliamentary democracy*. Basingstoke, Macmillan, 2000).

[1] For a comprehensive bibliography and selected book covers see the website on this book at: < http://www.afes-press-books.de/html/PDFs/vonBeyme_%20Bibliography_March2013.pdf> that will be occasionally updated.

K. von Beyme, *Klaus von Beyme*, SpringerBriefs on Pioneers in Science and Practice 14,
DOI: 10.1007/978-3-319-01535-4_2, © The Author(s) 2014

Das politische System Italiens. Stuttgart, Kohlammer 1970.

Die politische Elite in der Bundesrepublik Deutschland. München, Piper 1971; 2nd edition 1974.

V*om Faschismus zur Entwicklungsdiktatur. Machtelite und Opposition in Spanien*. München, Piper 1971.

Die politischen Theorien der Gegenwart. München, Piper 1972, 8th edition, Wiesbaden, Westdeutscher Verlag, 2000. (Span.: *Teorías politicas contemporaneas*. Madrid, Instituto des estudios políticos 1977). (Serbokroat.: *Suvremene politicke teorije*. Zagreb, Stvarnost 1975). Chines.: *Dangdai zhengzhi lilun*. Peking, Shangwu, 1990) (Polish: *Współczesne teorie polityczne*. Warsaw, Scholar, 2005).

Ökonomie und Politik im Sozialismus. Ein Vergleich der Entwicklung in den sozialistischen Ländern. München, Piper 1975, Paperback edition 1977. (Engl.: *Economics and Politics within Socialist Systems*. New York, Praeger 1982).

Gewerkschaften und Arbeitsbeziehungen in kapitalistischen Ländern. München, Piper 1977 (Engl.: Challenge to Power. Trade Unions and Industrial Relations in Capitalist Countries. London, Sage 1980).

Sozialismus oder Wohlfahrtsstaat? Sozialpolitik und Sozialstruktur der Sowjetunion im Systemvergleich. München, Piper 1977 (Short English version in: *International Political Science Review* 1981).

Das politische System der Bundesrepublik Deutschland. Wiesbaden, Westdeutscher Verlag, VS, 1979, 2010 11th edition (Engl.: *The Political System of the Federal Republic of Germany*. Aldershot, Gower/New York, St. Martins' 1983) (Sloven.: *Politični sistem Zvezne Republike Nemčije*. Koper, Visokošolsko središče, 2002).

Parteien in westlichen Demokratien. München, Piper 1982, 2nd edition 1984 (Engl.: *Political Parties in Western Democraties*. New York, St. Martins' Press, 1985). (Span.: *Los partidos políticos en las democracias occidentales*. Madrid, Siglo ventiuno 1986) (Ital.: *I partiti nelle democrazie occidentali*. Bologna, Zanichelli 1987).

Die Sowjetunion in der Weltpolitik. München, Piper, 1983, 2nd edition 1985 (Engl.: *The Soviet Union in World Politics*. New York, St. Martin's 1987).

Der Wiederaufbau. Architektur und Städtebaupolitik in beiden deutschen Staaten. München, Piper 1987.

Der Vergleich in der Politikwissenschaft. München, Piper 1988.

Hauptstadtsuche. Hauptstadtfunktionen im Interessenkonflikt zwischen Bonn und Berlin. Frankfurt, Suhrkamp 1991.

Theorie der Politik im 20. Jahrhundert. Von der Moderne zur Postmoderne. Frankfurt, Suhrkamp 1991, 3rd edition 1996. 4th edition 2007 (Span.: *Teoría politíca del siglo XX. De la modernidad a la postmodernidad*. Madrid, Alianza 1994)

Die politische Klasse im Parteienstaat. Frankfurt, Suhrkamp 1993, 2nd edition 1995 (Span.: *La clase política en el Estado de partídos*. Madrid, Alianza 1995). (Ital.: *Clase politica a partitocrazia*. Turin, UTET 1997).

Systemwechsel in Osteuropa. Frankfurt, Suhrkamp 1994 (Engl.: *Transition to Democracy in Eastern Europe*. London, MacMillan 1996) (Korean.: *talsahoezu ue wa zese joenwhan*. Seokang Dae Hak Kyoe Zalpanbu, 2000).

Der Gesetzgeber. Der Bundestag als Entscheidungszentrum. Opladen, Westdeutscher Verlag 1997 (Engl.: *Parliament as a Decision-Making Centre. The Case of Germany.* New York, St. Martin's 1998).

Kulturpolitik zwischen staatlicher Steuerung und Gesellschaftlicher Autonomie. Opladen, Westdeutscher Verlag 1998.

Die Kunst der Macht und die Gegenmacht der Kunst. Studien zum Spannungsverhältnis von Kunst und Politik. Frankfurt, Suhrkamp 1998.

Parteien im Wandel. Von den Volksparteien zu den professionalisierten Wählerparteien. Wiesbaden, Westdeutscher Verlag 2000. 2nd edition 2002 (Croat.: *Transformacija političkih stranaka*, Zagreb, Politička misao 2002).

Russland zwischen Anarchie und Autokratie. Wiesbaden, Westdeutscher 2001.

Politische Theorien in Russland 1789–1945. Wiesbaden, Westdeutscher 2001.

Politische Theorien im Zeitalter der Ideologien 1789–1949. Wiesbaden, Westdeutscher Verlag 2002.

Das Zeitalter der Avantgarden. Kunst und Gesellschaft. 1905–1955. Munich, Beck, 2005.

Föderalismus und regionales Bewusstsein. Ein internationaler Vergleich. Munich, Beck, 2007.

Die Faszination des Exotischen. Exotismus, Rassismus und Sexismus in der Kunst. München, Fink, 2008.

Geschichte der politischen Theorien in Deutschland 1300–2000. Wiesbaden, VS, 2009.

Vergleichende Politikwissenschaft. Wiesbaden, VS, 2010.

Kulturpolitik in Deutschland. Von der Staatskulturförderung zur Kreativwirtschaft. Wiesbaden, Springer VS, 2012.

Von der Postdemokratie zur Neodemokratie. Wiesbaden, Springer VS, 2013.

2.2 Edited and Co-edited Books in English

Theory and Politics. Festschrift für C.J. Friedrich. Den Haag, Nijhoff, 1971. (with Dante Germino): *The Open Society in Theory and Practice.* Den Haag, Nijhoff, 1974.

German Political Studies. London, Sage, Vol. 1, 1974.

German Political Systems (German Political Studies, Bd.2). London, Sage, 1976. (with Max Kaase): *Elections and Parties. Socio-political Change and Participation in the West German Federal Election of 1976* (German Political Studies, Bd.3). London, Sage, 1978. (with Hartmut Zimmermann): *Policy-Making in the German Democratic Republic* (German Political Studies, Bd. 5). Aldershot, Gower/New York, St. Martin's, 1984. (with Manfred G. Schmidt): *Policy and Politics in the Federal Republic of Germany* (German Political Studies, Bd. 6). Aldershot, Gower/New York, St. Martin's, 1985.

Right-wing Extremism in Western Europe. London, Frank Cass, 1988.

2.3 Articles in English

"Federal Theory and Party Reality in the Soviet Union". *Public Policy*, 1964: 395–412.

"Ministerial Responsibility and Cabinet Stability in European Parliamentary Systems". In: H. St. Commager et al. (eds.): *Festschrift für Karl Löwenstein*. Tübingen, 1971: 51–70.

"Authoritarian Regimes—Developing Open Societies?" In: D. Germino/K. von Beyme (eds.): *The Open Society in Theory and Practice*. Den Haag, 1974:109–120.

"Bibliographical Essay on the State of Research in Political Science in the Federal Republic of Germany". *German Political Studies*, London, 1974, vol. 1: 253–284.

"A Comparative View of Democratic Centralism". *Government and Opposition*, 1975, No. 3: 259–277.

"Basic Trends in the Development of the Functions of Parliament in Western Europe". In: European Parliament (ed.): *European Integration and the Future of Parliaments in Europe*. Luxembourg, 1975: 11–22. (with Ghita Ionescu): "The Politics of Employment Policy in Germany and Great Britain". *Government and Opposition*, 1977, No. 1: 88–107.

"The Politics of Limited Pluralism? The Case of West Germany". *Government and Opposition*, 1978, No. 3: 265–287.

"The Changing Relations between Trade Unions and the Social Democratic Party in West Germany". *Government and Opposition*, 1978: 399–415.

"Power and Trade Unions". In: *Power in Industrial Democracies. International Symposium*, Paris. International Foundation for Human Sciences. Paris, 1978: 143–168.

"Soviet Foreign Policy in Asia". In: *Journal of National Security*, Seoul, 1979, Vol. 9: 109–126.

"Political Culture and Electoral Behavior in West Germany". *German Studies Review*, 1980, No. 3: 415–433.

"Soviet Social Policy in Comparative Perspective". *International Political Science Review*, 1981, No. 1: 73–94.

"Détente and East–West Economic Relations". *The Journal of Politics*, 1981: 1192–1206.

"Modern Schools of Politics: Western Germany". *Government and Opposition*, 1982, No. 1: 94–107. (Reprinted in: W.G. Andrews (Hrsg.): *International Handbook of Political Science*. Westport/Conn., Greenwood, 1982:169–176).

"Elite Input and Policy Output. The Case of Germany". In: Moshe Czudnowski (ed.): *Does Who Governs Matter? Elite Circulation in Contemporary Societies*. DeKalb, Northern Illinois University Press, 1982: 55–67.

"Governments, Parliaments, and the Strucure of Power in Political Parties". In: H. Daalder/P. Mair (eds.): *Western European Party Systems. Continuity and Change*. London, Sage, 1983: 341–368.

"Neo-Corporatism. A New Nut in an Old Shell?" *International Political Science Review*, 1983: 173–196.

"Religious Pluralism and Political Behaviour in Europe". In: Giovanni Barberini (ed.): *Il pluralismo religioso in Europa fra culture e sistemi politici*. Rimini, Maggioli, 1983: 47–61.

"Coalition Government in Western Germany". In: Vernon Bogdanor (ed.): *Coalition Government in Western Europe*. London, Heinemann, 1983: 16–37.

"Policy Output in the GDR in Comparative Perspective". In: Klaus von Beyme/Hartmut Zimmermann (eds.): *Policy-Making in the German Democratic Republic*. Aldershot, Gower/New York, St. Martin's, 1984: 301–314.

"Do parties matter? The impact of Parties on the Key Decisions in the Political System". *Government and Opposition*, 1984, No. 1: 5–29.

"Kemalism in Western and Marxist Theory of Development". In: *Türkiye is Bankasi International Symposium on Atatürk*. Türkive is Bankasi Cultural Publications, 1984: 285–307.

"The Power Structure in the Federal Republic of Germany". In: Charles Burdick u.a. (eds.): *Contemporary Germany: Politics and Culture*. Boulder, Westview Press, 1984: 77–102.

"Unions and Multinational Companies". In: B. Wilpert/A. Sorge (eds.): *International Perspectives on Organizational Democracy*. London, John Wiley, 1984: 115–135.

"West Germany: Federalism". *International Political Science Review*, 1984: 381–396.

"The Role of the State and the Growth of Government". *International Political Science Review*, 1985, No. 1.: 11–34.

"Karl Marx and Party Theory". In: *Government and Opposition*, 1985, No. 1: 70–87.

"Policy-Making in the Federal Republic of Germany: A Systematic Introduction". In: Klaus von Beyme/Manfred G. Schmidt (eds.): *Policy and Politics in the Federal Republic of Germany*. Aldershot, Gower/New York, St. Martin's, 1985: 1–26.

"The Role of Deputies in West Germany". In: Ezra N. Suleiman (ed.): *Parliaments and Parliamentarians in Democratic Politics*. New York, Holmes & Meier, 1986: 154–175.

"The Challenges to Democracy: Corporatism, Social Movements and Interest Groups". In: *Interdisciplinary Science Reviews*, 1986, No. 2: 132–135.

"Attitudes of German Youth Toward Relations between the Two German States". In: *West Germany, East Germany and the German Question*. Baltimore, American Institute for Contemporary German Studies, The Johns Hopkins University, 1986: 33–43.

"Economic Relations as an Instrument of Soviet Hegemony over Eastern Europe?" In: Hans-Hermann Höhmann u.a. (eds.): *Economics and Politics in the USSR. Problems of Interdependence*. Boulder, Westview, 1986: 214–231.

"The Contemporary Relevance of the Concept of the State". *International Political Science Review*, 1986, No. 2: 115–119.

"Plea for Policy Analysis". *Policy Studies Journal*, 1986, No. 4: 540–544.
"National Consciousness and Nationalism: The Case of the Two Germanies". In: *Canadian Review of Studies in Nationalism,* 1986, No. 2: 227–248.
"The University and the Formation of Elites". In: Hermann Röhrs (ed.): *Tradition and Reform of the University under an International Perspective*. Frankfurt, Lang, 1987: 73–88.
"Parliamentary Oppositions in Europe". In: Eva Kolinsky (ed.): *Opposition in Western Europe*. London, Croom Helm, 1987: 31–51.
"Competitive Party Systems". In: Vernon Bogdanor (ed.): *Blackwell Encyclopaedia of Political Institutions*. Oxford, Blackwell, 1987: 123–127.
"Conservative Parties". In: Vernon Bogdanor (ed.): *Blackwell Encyclopaedia of Political Institutions*. Oxford, Blackwell, 1987: 135–137.
"Functions of Political Parties". In: Vernon Bogdanor (ed.): *Blackwell Encyclopaedia of Political Institutions*. Oxford, Blackwell, 1987: 412–416.
"Economics and Politics in a Socialist Country: Gorbachev's New Concepts". *Government and Opposition*, 1988: 167–185.
"Right-wing Extremism in Postwar Europe". *West European Politics*, 1988: 1–18.
"Perestroika: The Modernization of Socialism". *Politikon* (South Africa), Dec. 1988: 63–82.
"The Genesis of Constitutional Review in Parliamentary Systems". In: Landfried, Christine (ed.): *Constitutional Review and Legislation*. Baden–Baden, Nomos, 1988: 21–38.
"The Swedish Riksdag in a Comparative Perspective". In: Nils Stjernquist (ed.): *The Swedish Riksdag in an International Perspective*. Stockholm, Riksbankens Jubileumsfond, 1989: 13–23.
"Reform and Regional Policies in the Soviet Union". *The Journal of Communist Studies*, 1989: 267–284.
"Reconstruction in the German Democratic Republic". In: Jeffry M. Diefendorf (ed.): *Rebuilding Europe's Bombed Cities*. London, Macmillan, 1990: 190–208.
"Transition to Democracy—or Anschluss? The Two Germanies and Europe". *Government and Opposition*, 1990, No. 2: 170–190.
"Economic and Social Policy in the Bundestag". In: Uwe Thaysen et al. eds: *The U.S. Congress and the German Bundestag*. Boulder, Westview, 1990: 357–379.
"Architecture and Democracy in the Federal Republic of Germany". In: *International Political Science Review*, 1991, No. 2: 137–147.
"Post-modernity, Postmaterialism and Political Theory". In: Karlheinz Reif/ Ronald Inglehart (eds.): *Eurobarometer: The Dynamics of European Public Opinion*. London, Macmillan, 1991: 309–333.
"German Political Science: The State of the Art". *European Journal of Political Research*, 1991: 263–278.
"The Modernization of Socialism: Glasnost versus Perestroika?" In: Dankwart A. Rustow/Kenneth Paul Erikson (eds.): *Comparative Political Dynamics. Global Research Perspectives*. New York, Harper Collins, 1991: 262–291.
"The Legitimation of German Unification between National and Democratic Principles". *German Politics and Society*, 1991: 1–16.

"Electoral Unification: The First German Election in December 1990". *Government and Opposition*, 1991: 167–184.

"Inter-republican Economic Interactions in the Soviet Union". In: Alastair McAuley (ed.): *Soviet Federalism, Nationalism and Economic Decentralisation.* Leicester/-London, Leicester University Press, 1991:89–109.

"German political science: the state of the art". *European Journal of Political Research*, 1991: 263–278.

"The effects of Reunification on German Democracy: A Preliminary Evaluation of a Great Social Experiment". *Government and Opposition*, 1992: 158–176.

"The New National State and its Role in International Relations". In: Carl C. Hodge/-Cathal J. Nolan (eds.): *Shepherd of Democracy? American and Germany in the* 20*th Century*. Westport, Greenwood, 1992: 173–186.

"Political Mobilization and Participation". In: Meinolf Dierkes/Bernd Biervert (eds.). *European Social in Transition.* Frankfurt, Campus/Boulder, Westview, 1992: 371–390.

"West Germany and the New Germany: Centralization, Expanding, Pluralism, and the New Challenges". In: Clive S. Thomas (ed.): *First World Interest Groups. A Comparative Perspective*. Westport/Conn. Greenwood, 1993: 165–184.

"Parliamentary Crisis and how to Strengthen Democracy". In: Douglas Greenberg et al. (eds.): *Constitutionalism and Democracy. Transitions in the Contemporary World.* Oxford, Oxford University Press, 1993: 275–286.

"Germany". In: *An Australian Republic. The Report of the Republic Advisory Committee*. Canberra, Commonwealth Government Printer, 1993: 53–72.

"Regime Transition and Recruitment of Elites in Eastern Europe". *Governance*, 1993, No. 3: 409–425.

"Approaches to a Theory of the Transformation to Democracy and Market Society". *Transitions*, 1993, No. 2: 5–25.

"The head of the state: from monarchy to republic in the light of German and European Experience". In: George Winterton (ed.): *We, the people. Australian Republican Government*. Sydney, Allen & Unwin, 1994: 189–199.

"Approaches to a theory of transformation to democracy and market society". In: Geraint Parry (ed.): *Politics in an Interdependent World. Essays presented to Ghita Ionescu*. Aldershot, Edgar Elgar, 1994: 126–145.

"The Last Ideology of the old Intelligentsia". In: Adolf Bibic/Gigi Graziano (eds.): *Civil Society, Political Society, Democracy*. Ljubljana, The Slovenian Political Science Association, 1994: 95–109.

"A United Germany Preparing for the 1994 Elections". In: *Government and Opposition*, 1994, No. 4: 445–460.

"The Collapse of Socialism and its consequences for social science theory". *Law and State*. Vol. 49/50, 1994: 76–93.

"The Concept of Political Class. A New Dimension for Research on Elites?" In: Carlo Mongardi (ed.): *Gaetano Mosca, scienza politica e regime rappresentativo nell`età contemporanea*. Rom, Bulzoni 1995: 153–177.

"Democratic Transition in Central Eastern Europe". In: Mario Telo (ed.): *Democratie et construction européenne*. Brüssel, Editions de l'université de Bruxelles. 1995: 227–236.

"Biopolitical Ideologies and their Impact on the New Social Movements". In: Agnes Heller/Sonja Punscher-Riekmann (eds.): *Biopolitics: The Politics of the Body, Race and Nature*. Aldershot, Avebury 1995: 113–132.

"Party Leadership and Change in Party Systems: Towards a Postmodern Party State?" *Government and Opposition*, 1996, No. 2: 135–159.

"The Concept of Political Class: A New Dimension of Research on Elites?" *West European Politics*, 1996, No. 1: 68–87.

"A New Movement in an Ideological Vacuum: Nationalism in Eastern Europe". In: Gregory Andrusz et al. (eds.): *Cities after Socialism. Urban and Regional Change and Conflict in Post-Socialist Societies*. Oxford, Blackwell, 1996: 268–285.

"Political Theory: Empirical Political Theory". In: Robert E. Goodin/Hans-Dieter Klingemann (eds.): *A New Handbook of Political Science*. Oxford University Press, 1996: 519–530.

"Carl J. Friedrich. A founding father of comparative politics". In: Hans Daalder (ed.): *Comparative European Politics: The Story of a Profession*. London, Pinter 1997: 7–14.

"Walking on two legs: comparative politics in East and West". In: Daalder, *ibidem*. 1997: 215–226.

"Interest Groups in Parliamentary decisions in the German Bundestag". In: *Government and Opposition* 33, No. 1, 1998: 38–55

"The concept of totalitarianism—a reassessment after the end of Communist Rule". In: Achim Siegel ed.): *The Totalitarian Paradigm after the End of Communism*. Poznán Studies, Rodopi 1998: 39–54.

"Shifting National Identities. The Case of German History". In: *National Identities*, Vol. 1, No. 1 1999: 39–52.

"Democracy as Civil Society: the Mediating Structures". In: *Democracy. Some acute questions*. The Proceedings of the Fourth Plenary session of the Pontifical Academy of Social Sciences 22–25 April 1998. Vatican City, Pontificae academiae scientiarum socialium acta 1999: 229–247.

"Redefining European Security. The Role of German Foreign Policy". In: Carl C. Hodge (ed.): *Redefining European Security*. New York, Garland 1999: 165–179.

"Does the Constitution Need Reforming?" In: Olof Petersson et al. (eds.): *Democracy the Swedish Way. Report from the Democratic Audit of Sweden* 1999. Stockholm, SNS Förlag 1999: 17–46.

"The Bundestag—Still the Centre of Decision-Making?" In: Ludger Helms (ed.): *Institutions and Institutional Change in the Federal Republic of Germany*. Basingstoke, Macmillan 2000: 32–47.

"Fischer's move towards a European Constitution". In: Christian Joerges/Yves Mény/J. H. H. Weiler (eds.): *What Kind of Constitution for What Kind of Polity?* Florence, Robert Schuman Centre for Advanced Studies at the European University Institute/Harvard Law School, 2000: 61–84.

"Federalism in Russia". In: Ute Wachendorfer-Schmidt (Hrsg.) *Federalism and Political Performance*. London, Routledge, 2000: 23–39.

"Citizenship and the European Union". In: Klaus Eder/Bernhard Giesen (ed.): *European Citizenship*. Oxford, Oxford University Press, 2001: 61–85.

"A Future for Democracy in Russia?" In: Roland Axtmann (ed.): *Balancing Democracy*. London, Continuum, 2001: 214–237.

"Institutional Engineering and Transition to Democracy". In: Jan Zielonka (Hrsg): *Democratic Consolidation in Eastern Europe*. Vol. 1. Oxford, Oxford University Press, 2001: 3–24.

"Parties in the process of consolidation in East-Central Europe". In: Geoffrey Pridham/Attila Ágh (eds.): *Prospects for democratic consolidation in East-Central Europe*. Manchester, Manchester University Press, 2001: 138–156.

"Elite Relations in Germany". In: *German Politics*. Bd. 10, Nr. 2, 2001: 19–36. Reprint in: Stephen Padgett/Thomas Poguntke (eds.): *Continuity and Change in German Politics. A Festschrift for Gordon Smith*. London, Frank Cass, 2001: 19–36.

"Revolutions of 1989–90 in Eastern Europe". In: *International Encyclopedia of the Social and Behavioral Sciences*, Vol. 20. Amsterdam 2001: 13306–13310.

"Culture in Politics and Governance—European Experiences". In: Indra de Soysa/Peter Zervakis (ed.): *Does Culture Matter? The Relevance of Culture in Politics and Governance in the Euro- Mediterranean Zone*. Bonn, Zentrum für Europäische Integrationsforschung 2002: 31–36.

"The German Constitutional Court in an Uneasy Triangle between Parliament, Government and the Federal Laender". In: Wojciech Sadurski (ed.): *Constitutional Justice, East and West*. Den Haag, Kluwer 2002: 101–118.

"The Russian Constitutional Court in an Uneasy Triangle between President, Parliament and Regions". In: Wojciech Sadurski (ed.): *Constitutional Justice, East and West*. Den Haag, Kluwer 2002: 309–325.

"Constitutional Engineering in Central and Eastern Europe". In: Stephen White et al. (eds.): *Developments in Central and East European Politics 3*. Houndsmill, Palgrave Macmillan 2003: 190–210.

"Asymmetric federalism between globalization and regionalization". In: *Journal of European Public Policy*. Special issue: Alexander H. Trechsel (ed.): *Towards a Federal Europe?* Milton Park, Routledge Journals, 2005: 432–447. In a Book: Milton Park, Routledge, 2006: 32–47.

"The German Constitutional Court—A Model for New Democracies?" In: Ozan Ergül (ed.) *Democracy and the Judiciary*. Ankara, Union of Turkish Bar Associations, 2005: 67–100.

"Political Institutions—Old and New". In: R.A. W. Rhodes/Sarah A. Binder/Bert A. Rockman (eds.): *The Oxford Handbook of Political Institutions*. Oxford, Oxford University Press, 2006: 743–757; 2nd edition. 2008.

"Populism and Right Wing Extremism in Modern Democracies". In: Václav Peisertová/Vlad'ka Votavová (eds.): *Populism in Central Europe*. Prag, Association for International Affairs (AMO), 2007: 26–40.

"The evolution of comparative politics". In: Daniele Caramani (ed.): *Comparative Politics*. Oxford, Oxford University Press, 2008: 27–43.

"Federalism, Democracy and the Politics of Identity". In: Mohammad-Saïd Darviche/William Genieys (eds): *Multinational State-Building*. Montpellier, Pôle SuD, 2008: 27–42.

"The Historical Development of Comparative Politics". *Zeitschrift für Vergleichende Politikwissenschaft*, 2010: S 1–S 15.

"The evolution of comparative politics". In: Daniele Caramani (ed.): *Comparative Politics*. Oxford, Oxford University Press, 2nd edition. 2011: 23–36.

"Representative democracy and the populist temptation". In: Alonso, Sonia/Keane, John/Merkel, Wolfgang (eds.): *The Future of Representative Democracy*. Cambridge, Cambridge University Press, 2011: 50–73.

"Why is there no Political Science of the Arts?" In: Udo Hebel/Christoph Wagner (eds.): *Pictoral Cultures and Political Iconographies. Approaches, Perspectives, Case Studies from Europe and America*. Berlin, de Gruyter, 2011: 13–32.

"Difference and Indifference between Art Avant-Gardes and Democratic Politics". In: Kolja Raube/Annika Sattler (eds.): *Difference and Democracy*. Frankfurt, Campus, 2011: 369–376.

"Dimensions of Russia: developments after the USSR". In: Per-Arne Bodin/Stefan Hedlund/Elen Namli (eds.): *Power and Legitimacy—Challenges from Russia*. Milton Park/New York, Routledge, 2013: 8–27.

Part II
Key Texts by Klaus von Beyme on Comparative Politics

Chapter 3
Empirical Political Theory

Empirical political science divides into two main-streams.[1] The Weberian tradition is interested primarily in a reconstruction of social reality in a historical perspective and works *ex post facto* with typologies and ideal types. The Durkheimian tradition, deeply affected by French positivism after the fashion of Comte, takes as its motto *savoir pour prévoir* and is interested primarily in modeling reality by isolating dependent and independent variables.

This Durkheimian style of empirical political theory, in particular, supposes that models "should be tested primarily by the accuracy of their predictions rather than by the reality of their assumptions" (Downs 1957: 21). It, accordingly, has been particularly embarrassed by political science's failure to predict any major political events since 1945. The student rebellions of the 1960s, the rise of new fundamentalism, the collapse of communism, the peaceful revolution of 1989—all came as a surprise to political scientists.

Political science takes little comfort, either, in new tendencies in the natural sciences. Abandoning the old Baconian optimism that science does battle against ideology and superstition in the service of truth and utility, natural scientists influenced by autopoietic systems theory and chaos scenarios have given up on the idea of predicting major events on the macro level (Maturana 1985). Many social scientists have belatedly come to think similarly that macro-theoretical predictions are little more than 'informed guesswork'. The evolution of events can be reconstructed only *ex post facto*, and the task of theory is to keep open various options (Luhmann 1981: 157).

Political scientists face further systematic distortion of theory-building peculiar to their own field. While the positivistic mainstream endlessly echoes Max Weber's plea for value-free science, developments in the history and philosophy of science undermine many tacit assumptions of that model. In its mature

[1] This text was first published by Klaus von Beyme as "Empirical Political Theory", in: Robert E. Goodin/Hans-Dieter Klingemann (eds): *A New Handbook of Political Science*. Oxford, Oxford University Press, 1996: 519–530. By permission of Oxford University Press (<www.oup.com>) that was granted on 13 May 2013 by Ms. Mary Bergin-Cartwright.

K. von Beyme, *Klaus von Beyme*, SpringerBriefs on Pioneers in Science and Practice 14,
DOI: 10.1007/978-3-319-01535-4_3, © The Author(s) 2014

stage, social theory is increasingly subjected to 'social and political imperatives' of society. Broader social aims and interests, more than impetus behind scientific research (Barnes 1992). In spite of continuous de-ideologization of theory-building in the 1980s, social science theories inevitably start from the social conditions embedded in structuring political discourses (Wagner et al. 1991: 77; Wagner/Wittrock 1993).

3.1 A Chronology of Shifting Paradigms

Since the Second World War, there have been major shifts in the importance and focus of political theory driven largely from within the discipline itself. With the rise of the behavioral persuasion, itself disinterested in the great questions, came in the 1950s and 1960s a perceived decline in normative political theory (Miller 1990). Repudiating earlier metaphysical theories of the state, individual and groups became the starting-point of analysis. Theories such as Bentley's (1949) were revived. Positive political theory tried to restrict itself to conceptual analysis, and normative revolts remained isolated. In the late 1960s, however, there came a revival of 'grand' political theory, frequently in ideologized form. But since the late 1970s the great debates between 'positivists' and 'Marxists' have been exhausted and in their place *policy analysis* has emerged as the middle level of a new theory-building exercise embracing both empirical and normative elements (von Beyme 2006: 248 ff).

Major shifts in political theory have also arisen from extra-scientific political factors as well as from internal developments within the discipline. In most countries there was, over the postwar period, a decline of faith in the steering capacities of the political center. The 1960s and early 1970s were, especially in Europe, a time of planning illusions and Keynesian trust in anti-cyclical steering of the economy. In the 1970s, however, mainstream political science turned away from implementing grand ideological visions and toward empirical political studies, satisfying itself with typologies of policy cycles and political and societal actors. Theory-building in that period concentrated on variations in steering via social co-operation:

- *consociationalism* (Lijphart 1977);
- *neo-corporatism* (Schmitter 1981);
- *societal co-operation* (Willke 1983);
- *generalized political exchange* (Marin 1990);
- *private interest group government* (Streeck/Schmitter 1985), growing out of liberal-corporatist ideas, in part in resistance to attempts at "bringing the state back in" (Skocpol 1979); and
- models of the state as the steering center of society, as in *political cybernetics* (Deutsch 1966) and the *active society* (Etzioni 1968).

The 1980s experienced a new type of social and political actor, the new social movements. Schmitter dubbed this counterforce 'syndicalism': although a misnomer outside Latin countries, he still believed neo-corporatism to represent the best protection against unruliness and ungovernability (Schmitter 1981). In the 1980s the ecological problem, in particular, entered the agenda of political theory. Most Anglo-Saxon political scientists hesitated to construct more than partial theories on the basis of new social movements (Goodin 1992). However, philosophically minded European thinkers such as Beck (1986) hypothesized a 'risk society' with quite different dynamics than classical industrial society. While resisting the temptation of the autopoietic bandwagon, Beck insists (independently of the Frankfurt School) on the necessity of completing modernity by adding a new non-technocratic and non-rationalistic component. In the 1980s, Habermas himself abandoned his late-Hegelian historicist project of reconstructing ever more typologies of crises; and while his work on discourse (1987) holds out feint hope that new movements would succeed in the defending the life-world (*Lebenswelt*) against the 'system' with its alienating forces (bureaucratization, commercialization, justicialization), that hope seems resurgent in his more recent philosophy of the legal state (Habermas 1: 992).

Marxism withered away long before the collapse of 'real socialism'. Brilliant intellectuals such as Przeworski and Elster, building bridges from a variant of democratic socialism to rational choice theory, were dubbed 'rational choice Marxists'. Meanwhile, America experienced an unknown politicization of theory-building in the name of 'political correctness' and 'affirmative action' for underprivileged racial, ethnic and gender groups in society; even in the days of the 'Caucus' within the American Political Science Association, a new political and normative thrust had never had such an impact as in the 1980s (Ricci 1983: 188–190). Despite the growing interest in green issues and new social movements, Europe had apparently exhausted its desire for politicization in the 1960s and 1970s. All this time Marxism dominated the debate and set the agenda, even for 'bourgeois' thought, foregrounding issues of emancipation and participation in a neo-Rousseauean wave of radicalism. This basic conflict had contributed to the internationalization of the debate. As soon as that latent intellectual civil war in political theory disappeared, a new trend towards regionalization of paradigms became apparent.

The 1980s also saw a decline of neo-conservative thought, parallel to the withering away of the leftists' paradigms which had provoked that conservative backlash. Neo-liberalism became the predominant conservative mood in many countries. The liberals, the main targets of the communitarians, 'liberalized' in turn with fading counter-forces from the socialist camp and Marxist thought. Liberalism was able to turn back towards the ideal of a civil society, which became a basic consensus of enlightened democracies. They continued to emphasize the notion *l'homme* more than the participating *citoyen*. But they became more tolerant in turn toward political participation on the level of subsystems of the social system in terms of groups and new social movements. The peaceful velvet revolution in Eastern Europe showed the liberals that not all collective participatory democratic activity is bound to end up in a new authoritarian statehood (Cohen/Arato 1992).

3.2 The Geography of Paradigm Shifts

From American surveys, one would infer there is a substantial uniformity within political theory worldwide. Galston's (1993) APSA overview registers hardly any European contributions, apart from Habermas and a few French postmodernists. Despite the artificial uniformity of the debate as presented from the American perspective (which hardly takes cognizance of a foreign book unless it is translated), there is actually a growing diversification in political theory on the macro-normative level, whereas on the level of partial theory relevant for empirical studies uniformity is growing.

In the late 1970s and in the 1980s new divergences of major national cultures also had an impact on social and political theory. Galtung (1983) half-seriously offered a typology of intellectual styles which accumulated further evidence in its favor throughout the decade:

- *The French style*, which is preoccupied with language and art in social theory and which retains a stubborn institutionalism absent until recently from the non-Francophone mainstream, influenced empirical political theory primarily via postmodern thought (Lilla 1994):
- *The Teutonic style*, which Galtung had in 1983 lumped together with Marxism and thus associated with all then-socialist countries, has now abandoned the Marxist track and turned to the political right. But autopoietic theory, especially Luhmann (1984) and the Bielefeld school (Willke 1983), is as abstract and as far from operationalization as Teutonic reasoning should be. Despite tendencies toward orthodoxy, Luhmann (1984), like most postmodernists, shows no dogmatic zeal and usually merely ridicules theoretical adversaries as being 'old European ontologists'. The German debate is highly influenced by Bielefeld agnosticism towards any possibility of political steering, much less changing the world. Actors' theories are ridiculed, and in many respects political theory in Germany has come close to abandoning the actor's perspective altogether. Nevertheless, political science as a whole will cling to the possibility of tracing actors and their impact in the political process as a fundamental premise for certain types of study. A constructivist theory of science will facilitate this kind of 'philosophy as if'.
- There has never been a single *Anglo-Saxon style* of theorizing, as Galtung suggests. There were of course certain similarities between Britain and the USA, which were normally summarized under the rubric 'pragmatic'. But pragmatism as a philosophical dogma had less influence in Britain than in America. Positive political theory as an axiomatic, deductive type of theorizing (Riker/Ordeshook 1973: xi) had few followers in Britain.

3.3 Theory and Method: Levels of Theoretical Analysis

Political theory is typically done with scant attention to methodological issues. A division of labor has grown up, according to which theoreticians are absolved of responsibility for operationalizing their propositions and empiricists, in turn, are absolved of responsibility for confronting theoretical issues and are allowed instead to treat methodological questions merely as matters of research technique. But theory without methodological framework is sterile, and only very abstract approaches (such as dialectical criticism or autopoietic systems theory) literally identify theory and method. The complete identification of theory and method is as detrimental to empirical work as is the complete separation of it. A balance—so far more common in sociology than political science—is necessary.

Approaches to theory-building can best be seen as a matrix which differentiates between macro-level and micro-level theories, on the one hand, and systems-based theories and actor-based theories on the other. This matrix appears as Fig. 3.1. Hardly anyone works at the polar extremes. Only theory-building which operates on a very abstract level (such as Luhmann's) completely scorns actor-based theories, and by the same token approaches which start strictly with the individual (such as the behavioralists') need to introduce certain collective notions at a higher level of reasoning (Fig. 3.1).

Between those polar extremes of autopoietic systems theory and orthodox behavioralism are many possible theoretical starting-points There is, however, a tendency toward *rapprochement* between the extremist positions. By way of abstraction and induction, individualistic approaches can end up with models as abstract as certain systems theories. Systems theories, in turn, can be deductively

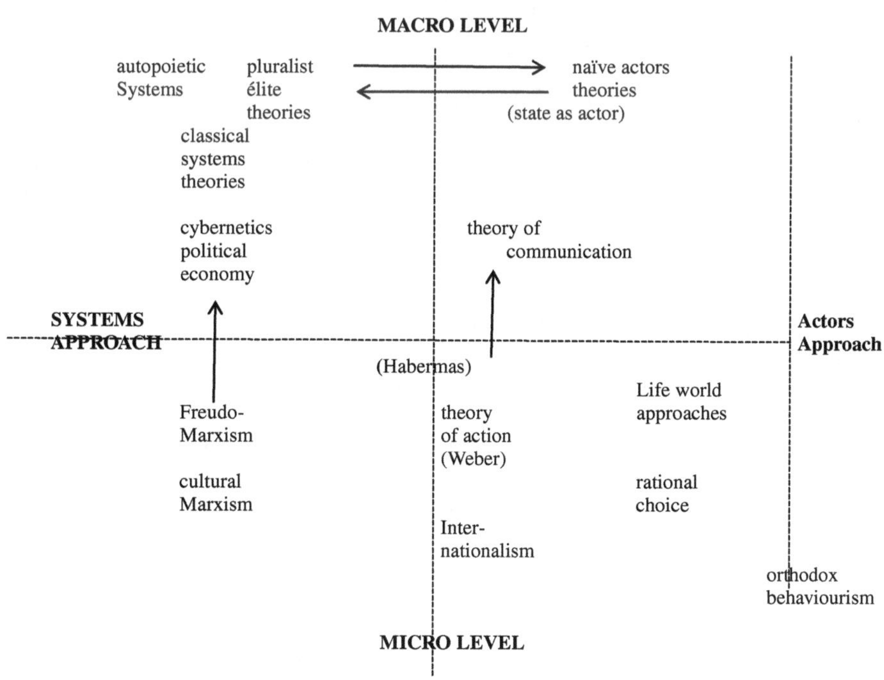

Fig. 3.1 Levels and approaches of theory building

differentiated into so many subfields and typologies of actors that they end up operating virtually at the micro-individualistic level of reasoning.

Political science thus gave up the classical modern notion of 'one discipline, one method, one preferred unit of analysis'. Mainstream political science lost faith in any fixed hierarchy of objects for research and methods of analysis. Political science theory-building predominantly takes place between the extreme poles of individualistic actor-based theories (such as interactionism and ethnomethodological approaches) and abstract system-based evolution. The meso-level, situated in between those micro and macro orientations, is essential for most questions of political science. Some notion of a collective actor is frequently adopted as an analytical device, even though every political scientist knows that institutions are not a unified entity of literally that sort.

The great recent success story in political science theory-building involves mathematical approaches, parallelling more mature social sciences such as economics. Rational choice modeling, in particular, conquered many U.S. departments and is now spreading through Europe. The APSA's account of the state of the discipline of political theory testifies to this triumph: in 1983, political theory consisted largely in a historical account of past empirical and normative political theories (Gunnell 1983), with Riker (1983) explicating some of his favorite coalition games as a mere aside; ten years later, 'formal rational choice' became a movement (Lalman et al. 1993: 77) and Riker was mentioned as an 'early contributor'. There seems now to be nothing more between 'formal rational choice theory' and 'normative political philosophy'.

How do we explain this astonishing success story? There are several reasons for it.

- The *neo-positivist claim for deductive political theory* is easiest to implement with formal models.
- *Rational choice approaches* can be applied to any behavior, from the most egoistic rationality to the most altruistic behavior of Mother Teresa, who also maximizes her strategy of helping the deprived.
- Political science, concentrated on the *meso-level between spheres of macro and micro theories,* needs to assume that-an actor-based approach is feasible. The actor of a rational choice approach is a construction which avoids questions about the real unity of a person.
- Rational choice *encourages quantification* and cumulative political science.
- Rational choice approaches were a *counter-balance against the dominance of behavioral studies* in earlier decades. It was easily combined with a multi-level analysis (especially in studies on the European Union) and with an enlightened neo-institutionalism, which spread in the 1980s (Scharpf 1989).

These advances of rational choice models point toward the stabilization of political science as a discipline. The development of theories in a discipline can hardly be the outcome of isolated individual predilections and insights. Only those theoretical approaches which comport with the internal rules of a scientific discipline, admitting of progressive elaboration by many hands, can form the basis for a theoretical mainstream.

3.4 Politial Trends and Their Impact on Theory-Building in the 1990s

A major shock to empirical political theory in the recent period came with the collapse of communism. That event not only forced revisions in theoretical explanations previously offered for developments in communist countries but also forced adaptations within the theoretical self-understandings of the victorious democracies themselves.

Most theories of modernization and the transition to democracy were modeled on developments in Southern Europe and South America in the 1970s (O'Donnell/ Schmitter 1986). Many of those developments were not comparable to the 1989 revolution, however. With those unprecedentedly simultaneous transformations of both economic and political systems, old assumptions of modernization theorists about economic prerequisites being essential for the success of political democratization were set on their heads (Karl/Schmitter 1991). The unique character of peaceful revolutions in 1989 even led to a testing of chaos theories taken from recent developments in biology and physics, although most of the applications of those theories remained merely metaphorical (cf. Marks 1992).

The First, Second and Third Worlds grew more similar after 1989. The decline of communism discredited theories such as Moore's (1966) of alternative roads to modernity. The authoritarian road to modernity followed by half-industrialized countries such as Italy and Germany collapsed in 1945. The totalitarian road to modernity primarily followed by predominantly agrarian countries such as China and Russia ended in 1989. Most transitional systems are democratizing, but it is unlikely that the final product of this process will be fully fledged democracy anytime in the forseeable future. More likely there will be a proliferation of *anocracies,* an admixture of anarchy and authoritarianism (Gurr 1991). Empirical political theorists have to confront the possibility of a certain degreee of backsliding among the consolidated democracies. The typological sequence of transitional societies—liberalization, democratization and consolidation was difficult to find in Eastern Europe, useful though this typology may have been elsewhere in the 1970s.

The great transformation in the early 1990s has been interpreted in terms of a crisis of modernization propeling us into a postmodern world (Baumann 1990). Many former Marxists have turned to some anarchical variation of themes of postmodernity and patchworks of minorities. It is unlikely, however, that a clear evolution from modernity to postmodernity will take place. Most reasonable postmodernists accept postmodernism only as a mere stage of modernity which implements its basic principles in a more consequential and systematic way than classical modernity. Insofar as it is not simply equated with post-materialism or with certain processes of differentiation and individualization which may lead to further decline of the old class social stratification and towards a development of life styles (Beck 1986), postmodernity is a set of theoretical assumptions rather than a clearly discernible new structure of society.

Some theoreticians in Europe (Beck 1993: 158) used the revolutions of 1989 as proof that system-based theories without actors were wrong. In some respects, however, those 'candle revolutions'—without revolutionary elites, ideologies, or mass organization—resemble more the 'evolution without subjects' hailed in Luhmann's theories. Certainly postmodern elements were present in those transformation processes, but a new scarcity makes it unlikely that a post-materialist and postmodern lifestyle will soon develop in Eastern Europe. On the contrary, even postmodernism in the West was shaken by events in Eastern Europe.

The claim that 'Communism perversion of modernity; post-communism enlightened postmodernity' is hardly tenable. Communism was already a hybrid of hyper-modernist megalomanic exaggerations of modernity, on the one hand, combined with pre-modern traits (implementing rationally planned systems through personalistic techniques of corruption, personal contacts and informal groups) on the other. In the West, postmodern theories emphasized a world of games. Lyotard's (1979) claim "let us quietly play" created an artificial world of games among parts of the Western intelligensia. This was possible only as long as it was protected by the Iron Curtain from major conflicts and intrusions; 1989 did away with this protection, and most of the postmodern problems are done away with by the more serious problems of survival.

Underlying the postmodernism debate is the search for a new balance between unity and plurality. The more successfully certain principles universalize, the more urgently components of plurality in modern societies emphasize their right to exist (Marquard 1987). This claim is normally realized by new social movements, but it has proven to be too early to see modernity on the road to a "society of movements" (Neidhardt/Rucht 1993). Empirical research suggests, rather, that the movements are vital for the first stages of the policy process—agenda-setting and policy formation—but that decision-making, implementation and evaluation are predominantly done by the traditional institutions and organized political forces, such as interest groups and parties.

It is also not by chance that the most recent normative debate between liberals and communitarians has now crossed the Atlantic and entered the vacuum left by the now-defunct Marxist ideological debates. A new nor- mative minimal consensus is developing. The paradox of the early 1990s is this. Empirically minded European political scientists are looking for new analytical tools in America. But what they find there is broad skepticism toward the old positivistic, behavioralistic paradigms. They are also discovering a new message, rather normative in character. Furthermore, it is a message which they are now ready to accept, since the older social-democratic consensus in the North European countries has withered away. The pragmatic left in Europe—deeply affected by the erosion of communism, even though they did not share its views—needed a new normative orientation.

Though many superficial observers tend to think of the development of paradigms in terms of cumulative progress, we are increasingly realizing that there are Kuhnian revolutions but not in the same sense as in the natural sciences (Kuhn 1970). There are revivals of old positions. Neo-Aristotelianism is not as dead as the pre-Copernican vision of the world. In political theory, we see a series of small innovations rather than big revolutions. Most of them are not created by established mainstream thinkers but by theoreticians who stand apart from mono-disciplinary research in the spirit of "creative marginality" (Dogan/Pahre 1990: 182 ff.).

References

Barnes, B., 1992: *T. S. Kuhn and Social Science* (New York: Columbia University Press).
Baumann, Zygmunt, 1990: "From Pillar to Post", in: *Marxism Today*: 20–25.
Beck, Ulrich, 1986: *Die Risikogesellschaft* (Frankfurt: Suhrkamp) [English edition, 1992: *Risk Society* (London: Sage)].
Beck, Ulrich, 1993: *Die Entdeckung des Politischen* (Frankfurt: Suhrkamp).
Bentley, Arthur, 1949: *The Process of Government*, New edition (Bloomington: Principia Press).
Beyme, Klaus von, [11]2006: *Die politischen Theorien der Gegenwart* (Wiesbaden: VS Verlag für Sozialwissenschaften).
Cohen, J. L., Arato, 1992: *Civil Society and Political Theory* (Cambridge: MIT Press).
Deutsch, Karl. W., 1966: *The Nerves of Government* (New York: Free Press).
Dogan, Mattei; Pahre, R., 1990: *Creative Marginality: Innovation at the Intersections of Social Sciences* (Boulder: Westview).
Downs, Anthony, 1957: *An Economic Theory of Democracy* (New York: Harper).
Etzioni, Amitai, 1968: *The Active Society* (New York: Free Press).
Galston, W., 1993: "Political Theory in the 1980s: Perplexity Admidst Diversity", in: Finifter A. (Ed.): *Political Science: The State of the Discipline II* (Washington, D.C.: American Political Science Association): 27–53.
Galtung, Johan, 1983: "Struktur, Kultur und intellektueller Stil", in: *Leviathan*, 11,3: 303–338.
Goodin, Robert. E. 1992: *Green Political Theory* (Oxford: Polity).
Gunnell, J. G., 1983: "Political Theory: The Evaluation of a Subfield", in: Finifter A. (Ed.): *Political Science: The State of the Discipline* (Washington, D.C.: American Political Science Association): 303–338.
Gurr, Tedd. R. 1991: "The Transformation of the Western State: The Growth of Democracy, Autocracy and State Power Since 1800", in: Inkeles A. (Ed.): *On Measuring Democracy: Its Consequences and Concomitants* (New Brunswick: Transaction Publishers).
Habermas, Jürgen, 1992: *The Philosophical Discourse of Modernity* (Cambridge: MIT Press).
Habermas, Jürgen, 1992: *Faktizität und Geltung. Beiträge zur Diskurstheorie des Rechts und des demokratis-Chen Rechtsstaats* (Frankfurt: Suhrkamp).
Karl, T., Schmitter, Philippe, 1991: "Modes of Transition in South Africa, Southern Europe and Eastern Europe", in: *International Social Science Journal*, 43: 269–284.
Kuhn, Thomas, S., [2]1970: *The Structure of Scientific Revolutions* (Chicago: University of Chicago Press).
Lalman, D. et al., 1993: "Formal Rational Choice Theory: A Cumulative Science of Politics", in: Finifter A. (Ed.): *Political Science: The State of the Discipline II* (Washington, D.C.: American Political Science Association): 77–103.
Lijphart, Arend, 1977: *Democracy in Plural Societies* (New Haven: Yale University Press).
Lilla, M., 1994: *New French Thought: Political Philosophy* (Princeton: Princeton University Press).
Luhmann, Niklas, 1981: *Politische Theorie im Wohlfahrtsstaat* (Munich: Olzog).
Luhmann, Niklas, 1984: *Soziale Systeme* (Frankfurt: Suhrkamp).
Lyotard, J.-F., 1979: *La Condition Postmoderne* (Paris: Minuit).
Marks, Garry, 1992: "Rational Sources of Chaos in Democratic Transition", in: *American Behavioral Scientist*, 35: 397–421.

Marin, Bernd (Ed.), 1990: *Generalized Political Exchange* (Boulder: Westview).
Marquard, Odo (Ed.), 1987: *Einheit und Vielfalt* (Hamburg: Meiner).
Maturana, H. R., [2]1985: *Erkennen. Die Organisation und Verkörperung von Wirklichkeit* (Brunswick: Vieweg).
Miller, D., 1990: "The Resurgence of Political Theory", in: *Political Studies*, 38: 421–437.
Moore, Barrington, 1966: *Social Origins of Dictatorship and Democracy* (Boston: Beacon Press).
Neidhardt, Friedhelm; Rucht, Dieter, 1993: "Auf dem Weg in die Bewegungsgesellschaft", in: *Soziale Welt*, 44,3: 305–326.
O'Donnell, Guillermo; Schmitter, Philippe (Eds.), 1986: *Transitions from Authoritarian Rule* (Baltimore: Johns Hopkins University Press).
Ricci, D. M., 1983: *The Tragedy of Political Science: Politics, Scholarship and Democracy* (New Haven: Yale University Press).
Riker, William H., 1983: "Political Theory and the Art of Heresthetics", in: Finifter A. (Ed.): *Political Science: The State of the Discipline* (Washington, D.C.: American Political Science Association): 47–68.
Riker, William. H.; Ordeshook, P. C., 1973: *An Introduction to Positive Political Theory* (Englewood Cliffs: Prentice-Hall).
Rucht, Dieter, (Ed.), 1991: *Research on Social Movements* (Frankfurt/Boulder, Colo.: Campus/Westview).
Scharpf, Fritz. W., 1989 : "Politische Steuerung und politische Institutionen", in: *Politische Vierteljahreschrift*, 30,1: 10–21.
Schmitter, Philippe. 1981: "Interest Intermediation and Regime Governability in Contemporary Western Europe and North America", in: Berger S. (Ed.): *Organizing Interests in Western Europe* (Cambridge: Cambridge University Press): 287–330.
Schwarzenbach, S. A., 1991: "Rawls, Hegel and Communitarianism", in: *Political Theory*, 9: 539–571.
Senghaas, Dieter, 1982: *Von Europa lernen* (Frankfurt: Suhrkamp).
Skocpol, T., 1979: "Bringing the State Back in: Strategies of Analysis in Current Research", in: Evans, P. R.; Rueschemeyer, D. R.; Skocpol, T. (Eds.): *Bringing the State Back In* (New York: Cambridge University Press): 3–37.
Streeck, Wolfgang; Schmitter, Philippe (Eds.), 1985: *Private Interest Government* (London: Sage).
Wagner, P.; Wittrock, B., 1993: *Social Sciences and Societal Developments: The Missing Perspective* (Berlin: Mimeo., WZB).
Wittrock, B.; Whitley, R. (Eds.), 1991: *Discourses on Society: The Shaping of the Social Science Disciplines* (Dordrecht: Kluwer).
Willke, H., 1983: *Entzauberung des Staates. Überlegungen zu einer Sozietalen Steuerungstheorie* (Königstein: Athenäum).

Chapter 4
The Evolution of Comparative Politics

4.1 Introduction

The evolution of comparative politics has been classified in stages, such as the 'pre-paradigmatic phase' which was not dominated by a single theoretical approach in the scientific community and the 'paradigmatic phase', in which the scientific community adhered to a dominant theory.[1] According to this classification (Chilcote 1994: 58) it is followed by as 'crisis phase' and finally ends up in a 'phase of scientific revolution' which occurs when the scientific community shifts to different paradigms. Kuhn's frequently abused term 'paradigm' is hardly applicable in this context. Most phases in the evolution of political science have been 'pre-paradigmatic' in the sense that no single approach predominated completely. Chilcote was aware that his subdivision of 'traditional', 'behavioural' and 'post-behavioural' approaches do not precisely fit Kuhn's definition of paradigms. Only the dominance of the 'behavioral revolution' came close to the idea of a paradigm which conquered the community and tolerated deviant approaches only in marginal positions. But the typology is ethnocentric in so far as it generalizes the American development. In Europe there was never a dominance of the behavioral approach. This author prefers a threefold classification for the evolution of' comparative politics with stages such as 'pre-modern', 'modern' and 'post-modern'.

[1] This text was first published by Klaus von Beyme as "The Evolution of Comparative Politics". In: Daniel Caramani (ed.): *Comparative Politics*. Oxford University Press, 2008: 27–43. By permission of Oxford University Press (<www.oup.com>) that was granted on 13 May 2013 by Ms. Mary Bergin-Cartwright.

K. von Beyme, *Klaus von Beyme*, SpringerBriefs on Pioneers in Science and Practice 14, DOI: 10.1007/978-3-319-01535-4_4, © The Author(s) 2014

4.2 Comparisons in Pre-modern Times

The pre-modern stage or traditional approach to comparison since Aristotle was highly speculative and normative, mostly ethno-centric and used comparison in an anecdotic way, but hardly ever tried a systematic comparison over time. Political science is the youngest social science in terms of modern professional performance. Comparative political science owes a lot to other sciences: philosophy since Aristotle, to legal constitutionalism from Bodin to Bryce, political economy from Smith, Ricardo, Bentham, Marx and Mill. Mill was especially fruitful in methodology. In the nineteenth century sociology—a term coined by Auguste Comte—was added and soon became important to help political science to liberate itself from jurisprudence and to be transformed into a 'social science'.

In the era of Renaissance Machiavelli came close to a social science approach, minimizing the philosophical normativism of former times. Later comparisons were sometimes used to criticize one's regime, disguised under the description of distant systems, as in Montesquieu's *Lettres persanes* (1721) or even utopian constructions of systems such as in the *Oceana* (1656) of James Harrington. One of the earliest and most complete comparisons was the work of a follower of Machiavelli's, Boccalini (1614: 1). In his *Ragguagli di parnaso* a 'università de' politici' was summoned by Apollo on the Parnass and had to give responsible answers concerning their various political systems. In spite of many insights this work was distorted by a blind hatred against the 'imperialist power' of that time, the Spanish monarchy, which according to Boccalini interfered too much in Italian affairs. Many historical comparisons in early modern times—from Machiavelli's *Discourses on the First Ten Books of Titus Livius* (1513) to Montesquieu's *Considérations sur les causes de la grandeur des Romains et de leur décadence* (1734)—were rather a-historical confrontations of Roman experiences and modern life of states. Reasons for the decay of the Roman Empire were popular as a kind of normative warning for modern states. The diachronic comparisons treated various systems like contemporary societies. But they did not help to develop a critical methodology of comparison.

The evolutionist counter-reaction since the French revolution was also not favourable for a scientific theory of comparison. History in the nineteenth century turned increasingly to historicism and the discipline developed more reservations against the comparative method than former political theories in the age of enlightenment. Every historical event and development was declared 'unique'. Already Goethe once said 'only blockheads compare'. But he had only works of art and literature in mind. Goethe was afraid that mediocre connaisseurs might avoid a value judgement about works of arts. This was indeed a permanent danger of the comparative sciences in many fields: relativism describing various historical solutions neither led to a conclusion nor even to a prognosis about possible future historical developments. Otto Hintze in German history with his comparative typologies was an outsider in his discipline. Troeltsch (1922, 1961: 191), another social science-oriented historian, accepted comparisons only when they kept their 'methodological and heuristic character' at the level of building of hypotheses. This was conform with older pioneers of the historical method such as Droysen (1960: 163) who knew already that without implied comparisons no meaningful hypothesis could be found in an ocean of facts and motivations among historical actors. American history in the twentyth century with Charles Tilly, Barrington Moore, Theda Skocpol and many others was leading the anti-historicist counter-reaction and gave up the anti-comparative bias of historicist historiography.

Since Tocqueville (1961: 5, 12) it was a widespread assumption that a new world under conditions of modernisation needs a 'new political science', able to work on certain social developments which are likely to spread in all modern societies. The USA were only a pretext to denounce the threats of equalisation and democratisation which were likely to spread also in Europe. Even a pioneer of comparative methods such as Mill (1840, 1859: 62) in the *Edinburgh Review* resented that his friend Tocqueville in his seminal book "has bound up in one abstract idea the whole of tendencies of modern commercial society, and given them one name—democracy". This was an important precedent in the history of book reviews, criticizing remainders of a teleological approach to the evolution of comparative politics.

In pre-modern political theories certain features of the decision-making process in *polities* were mostly deducted as constant types, but hardly ever scientifically analysed as *politics*. The *policies* were still more rarely investigated. Some utopias, such as Harrington's *Oceana* (1656) represented a notable exception. The theory of the 'reason of state' of Machiavellians like Giovanni Botero (1589, 1948: 58ff)—which represented a kind of 'Jesuit welfare-Machiavellism'—went far beyond Machiavelli's obsession with foreign and military policies in so far as he developed the elements of domestic welfare policies in different states as the main criterion for political stability.

Pre-modern comparisons were mostly aiming at classifications of whole political orders (*polity*). Only in modern times *politics* was compared when the techniques of ruling in theories of 'reason of the state' or 'sovereignty' were discussed since Machiavelli. Comparisons were rather simple typologies, such as those counting the number of rulers. Many of them contained normative assumptions. The characteristic features were not always logically consistent, such as Montesquieu's classification of monarchy, republic and despotism. Voltaire already mocked at this typology which appeared to him as logic as the categories of a 'church registration of births', containing the elements: 'male', 'female' and 'illegitimate'.

Pre-modern approaches in the nineteenth century 'modernized' by turning away from static ontological classifications to historical theories of evolution. The most influential models were the evolutionary model of Darwin and the historical materialism of Marx with his five historical stages. In comparative social science the two extremes sometimes came to a synthesis as in the evolutionary model of Herbert Spencer. The three authors had a predominant approach to theory-building: Darwin's approach was a kind of early 'functionalism', Marx' adopted from Hegel the dialectical method, and Spencer established himself as a precursor of a 'system's theory'. Since Mill (1959: 253) the logic of social science implied two methods, the 'method of agreement' and the 'method of difference'. The first method was a kind of 'artificial experiment', the second method was to be applied in situations where experiments were unfeasible. Early comparisons in the pre-modern era were obsessed to find similarities. Only in the twentyth century the primacy of the 'method of difference' was increasingly developed.

4.3 Comparisons in the Work of Thinkers in Classical Modernity

4.3.1 Scientific Comparison as Controlled Experiment

There is no agreement when modernity starts. It is normally scheduled earlier in art and literature, including large parts of the nineteenth century. In the social sciences modernity is scheduled later. The criteria of definition—a truly scientific theory which can be controlled empirically—offers a more precise proof for modernity than works of arts. In order to avoid quarrels of definition this author chooses the term 'classical modernity' for the new social sciences in the twentyth century. It coincides largely with the establishment of separate disciplines in the social sciences, such as sociology and political science. The neighbouring social sciences, such as public law, political economy or general history, in the nineteenth century still pretended to deal with politics in the evolution of modernisation and specialisation. In the twentyth century they withdrew from comparative politics. 'Comparative economics' continued to exist as a subfield, but it never played the dominant role of comparative politics in political science because of the mathematical character of the most influential economic models.

The German *Staatslehre* (theory of state), located in the law departments, had a certain influence among the founding fathers of American Political Science, such as Francis Lieber, a Prussian refugee who taught at Columbia University. But the second generation of American scholars, including Woodrow Wilson, already abhorred general theories of the state. A pioneer of group theory, such as Bentley (1908, 1949: XIX), after studies in Berlin and Freiburg, turned away from the 'spooks in the grain fields' which he discovered in the metaphysical-minded *Staatslehre*. On the basis of American pragmatism he developed in Chicago an extremely anti-state theory of politics. The groups—instead of states—were now the basic concept of comparative analysis. Former deductive elements of comparisons were substituted by inductive observations. In Europe British guild socialism and Harold Laski created an equivalent of an empirical political theory, directed against the 'statism' of continental political theory.

Spencer among the 'evolutionists' of the nineteenth was the most influential thinker for empirical comparative sciences. But in spite of his variation of a system's theory the founder of structural-functional systems theory, Parsons (1961:

3), opened his theory of action with the attack: "Spencer is dead, but who killed him and how?". Spencer was not killed, but increasingly forgotten—as most of the theories of historical stages in the nineteenth century, especially Auguste Comte. Parsons tried to smooth down his verdict in a footnote: "Not, of course, that nothing in his thought will last. It is his social theory as a total structure that is dead". Critical Rationalism of the neo-positivist school of Karl Popper later was keen of 'hunting' what it called 'historicism'. Charles Merriam and other founders of political science in the USA were more tolerant and recognized at least a certain progress because this kind of historicism was 'historical-comparative' and no longer normative and purely deductive as the traditional mainly anecdotal comparisons in the classical political literature.

The various approaches had some assumptions in common, even when Durkheim, Weber and Pareto, for Parsons (1961) the champions of modern social science, differed in many ways. Oddly enough they did not relate to each other—though all the three were able to read the language of the two others. Historical factors, which Darwin (biological struggle for life) or Marx (class conflict grounded on the contradiction between the economic base and the political superstructure) saw as the driving force behind the evolution, were no longer accepted. But nevertheless a 'dominant variable' was behind the modern evolution: for Weber it was occidental rationalism and bureaucracy, for Durkheim it was the division of labour, leading to a kind of 'organic solidarity' and for Pareto it was the cycles of rise and decay of elites.

In spite of these differences there are four principles common to all thinkers of classical modernism:

1. *History is not identical with evolution.* There is no longer a 'telos', a final point where the evolution is aiming at. Insofar political theory has to give up the old normative idea: 'historia magistra vitae'—history can no longer serve as a teacher for later generations.
2. *Theory and practice are divided.* The scholar is not obliged to take political action as was still the case in the theories of Marx and Engels. This has not prevented Pareto and Weber once serving as a candidate for parliamentary election. Fortunately for the progress of social science they failed. Anti-normative value-free science was discovered as a protection against the interferences of the state as well as against the demands of political groups which consider scholars close to their ideologies. There is no longer the hope for a 'philosopher king' who combines knowledge and political action.
3. *The spheres of life and subsystems of the social system are separated and autonomous.* There is no hope that the political system—as in the times of absolutism—will be able to steer the subsystems of the whole society. With totalitarian ideologies this kind of hope was renewed, until the dictatorships collapsed between 1949 and 1989. Carl Schmitt was probably the most influential political theoretician who tried in a heroic turn back to 'revolutionary conservatism' to re-install 'political decision' in its primordial rights in society. The driving force behind this was that Schmitt was afraid of a lasting dominance

of the economy in society. Especially after the failures of dictatorship political theory reduced its claims and renounced of a primacy of the political subsystem.

4. Value free science should be *comparative and theory-guided*, not just an enumeration and typology of institutions as in the works of some 'pioneers of comparative government' as the new discipline was initially called from James Bryce to Carl J. Friedrich.

Typologies are the initial stage of theory-building in order to develop a hypothesis for empirical work. Some typologies pretended to be a theory. Theories contain generalizations about political reality, typologies are abstractions about political reality according to some formal criteria. Some theories were close to one approach, such as 'functionalism'. In other cases a methodology such as the behavioralist approach tried to gain the status of a monopolistic theory. In recent times 'rational choice' showed similar tendencies. Some typologies were comparative only in an indirect way. Even Max Weber's typology of types of legitimate rules are not free of remainders of the old debate 'uniqueness versus comparability' because the 'ideal types' contained elements of uniqueness as 'individual totalities'. The ideal types served to elaborate the special features of social institutions. Only in the time of classical modernity typologies of dynamic processes were offered. Historical research continued, however, to suspect such taxonomies as they were developed in Crane Brinton's *Anatomy of revolutions*.

We should not identify the modern breakthrough to scientific comparisons with behaviorism, as sometimes occurred in American literature. Even the broader and less rigid form of 'behavioralism'—which dominated for a while the torch-bearers of modern political science in the 1950s and 1960s—was soon combined with other approaches, such as functionalist system's theories. Functional considerations were not completely compatible with the strictly individualistic assumptions of behaviorism. This contradiction can be shown in the seminal research on *Civic Culture* by Almond and Verba (1963: 52, 68). The authors had some misgivings whether the uses of comparative survey studies would allow to discover the 'uniformity of a psychological type' in a whole country. In order not to distort the results it was recommended to concentrate on the behaviour or attitudes that are least determined by the structure of the situation. Behavioralism started from a rather mechanistic stimulus–response-model of behavior, functionalism was closer to organic models. System's theory is holistic and presupposes 'purposes' of a system which was speculative for many behavioralists. But neither system's theory nor behavioralism in practice stayed completely dogmatic, so that a cooperation of various methodological tenets was possible.

Beyond the general meta-theoretical consensus of all empirical scholars in political science behavioralism developed major tenets as Easton (1965: 7) classified them in his *Framework for Political Analysis*:

- Regularities or uniformities in political behaviour should be expressed in generalizations or theory.

- The validity of these generalizations has to be tested. Contrary to Popper's orthodoxy which admitted only 'falsification', 'verification' was possible.
- Techniques of seeking and interpreting data have to be developed.
- Quantification and measurement in the recording of data.
- Values as distinguished between propositions relating to ethical evaluation and those relating to empirical explanation.
- 'Pure science', or the seeking of understanding and explanation of behaviour before utilization of knowledge for solution of societal problems.
- Integration of political research with that of other social sciences.

Only when political science was well established in American universities 'comparative politics' developed its dominant position, in theory-building of the discipline as well as in the evaluation of the ranks of individual scholars in the scientific community. American and German rankings since Somit and Tanenhaus (1964: 66) and Falter and Klingemann (1998, 2006) have shown that comparative scholars were on top of the hierarchies of reputation among the colleagues of the discipline. In American political science after 1945 seven scholars among the top ten were in comparative politics, two in theory and one in international politics. In the German case there was only one scholar in the field of international relations among the top ten in political science. All the rest—mostly Americans, one Italian, one Dutch, one German—were scholars in comparative politics. Thanks to a new world situation, the development of the 'third world' and modernisation theories which dealt with the transition from traditional societies to modern democracy, the main theoretical innovations in political science were developed by comparative scholars.

Only political science has well accommodated the comparative aspects in a special subfield. The existence of a sub-discipline 'comparative politics', was, however, not uncontested in the organized discipline. There was never a dominant behavioral stage of comparative politics as postulated by Chilcote (1994: 56). From sociology—in methodological questions far more sophisticated—political science inherited two traditions:

1. *The historical-institutional tradition* of Max Weber was comparative. Weber almost excessively looked in all the great cultures of the world for comparable elements, especially in his sociology of religions.
2. The opposite school was initiated by Herbert Spencer and lead by Émile Durkheim (1950: 137) He opposed a subdiscipline of 'comparative sociology' because comparison for him was "la sociologie même" (sociology itself). System's theory with various degrees of intensity joined this opinion, most vehemently the autopoietic variant of Luhmann (1970: 25, 46) who tried to go beyond Talcott Parsons' structural–functional theory. In economics the former historical orientations were increasingly substituted by mathematical models. One branch of political science took over this kind of approach. Downs (1957: 21) gave the prognostic abilities of political theory priority over its capacity to describe the political reality: "Theoretical models should be tested primarily by

the accuracy of their predictions rather than by the reality of their assumptions." Comparative politics therefore has been particularly embarrassed by its failure to predict any major political events since 1945. The student rebellions of the 1960s, the oil crisis, the rise of new fundamentalism, the collapse of communism of 1989—all these events came as a surprise to comparative political scientists. Some forecasts were correct, such as the possible end of the Soviet Union, but the prognosis was based on the wrong reasons, such as a Chinese-Soviet war in the bestseller of Amalrik (1970). Forecasting had to lower its ambitions. In forecasting short-term electoral results the discipline boasts of only 5 % margins of error. Many political scientists came to accept that macro-theoretical predictions are little more than informed guess-work. The evolution can only be reconstructed *ex post facto*.

The 'most similar design' of comparison was still widespread, but some researchers preferred a "most different system design" (Przeworski/Teune 1970: 31ff). In historical perspective the similarities were mostly demonstrated by the diffusion of institutions, a method widely applied in anthropology. System's theory on the contrary looked for dissimilarities which were able to serve as 'functional equivalents' in various systems with sometimes 'similar results'. Only the post-modernist Luhmann went as far as to claim the one-party state in Communist systems was a functional equivalent of the pluralist democratic party regimes in the West.

4.3.2 Typologies and Classifications, The First Step to Comparison

There is a large consensus that the main purpose of comparative research is not comparing but explaining. Comparison is a tool for building empirically falsifiable explanatory theories. A first step is a rigid classification. 'Miscomparing' starts from 'misclassification' 'concept stretching' and what Sartori (1991: 248) called "degreeism". This neologism meant the replacement of dichotomous treatment by continuous notions. According to Sartori the classification by degrees leads to logical messiness. The result Sartori is afraid of could be called in an ironic way the discovery of non-existing 'dog-cats'. Pre-modern as well as some modern typologies do not live up to these rigid criteria. The predicaments of typological work as a base for comparison show that system's theories are frequently characterized by geometrical obsessions of order, institutional typologies on the other hand, rather follow arithmetic intentions. Religious remainders invade typologies, trinities are discovered everywhere (cf. scheme).

These trinities do not violate Sartori's verdict against "degreeism", but logically we would prefer dual typologies, such as Spencer's (society of warriors/industrial society), Durkheims (mechanical/organic solidarity), Tönnies (community/society), Bagehots (dignified/efficient parts of the constitution). If more than three elements are put into one taxonomy the danger cannot be excluded that the theoretical value is reduced to a kind of checklist. Remainders of a teleological typology are frequently found in trinitarian classifications. The third element is quite often hailed as the normatively desired type in the development.

The problem of different criteria in classifications of types of regimes was not always solved—not even in modern typologies. Probably the last scholar who tried to classify all the regimes in history—from anarchy and tribal rule to totalitarian dictatorships—was Friedrich (1963: 188f). He listed 13 types of rule, but they lacked a common criterion. Some were characteristic of early societies, others were only minor institutional variations of representative government, such as presidential or parliamentary systems. Some regimes, such as the types of dictatorship, were classified not by their institutional characteristics, but by the extent of control over the citizens. In formal sociology there was for a while a tendency to classify outrageously, as in the treatises of Georges Gurvitch in France or of Leopold von Wiese in Germany. In political science this happened only occasionally in the classification of regimes. Classifications of regime types should not reinvent Greek notions as Küchenhoff (1967) did, but rather try find a common-sense solution in terms which are accepted by scholars as well as by the public debate.

Excessive preoccupation with terminological clarity revealed a predicament: neo-logisms, mostly in Greek or Latin, reduced terminological ambiguities, but they had no chance to enter the public debate. Political science terminology is imbued with traditional perceptions of politics and can hardly proceed like chemistry or medicine in preserving purely scientific jargon. As an example for this predicament we might check the type of a 'semi-presidential system'. The term is not quite correct, because a system with a popularly elected president remains a variation of a parliamentary system: parliament can topple the government by votes of non-confidence and the president can dissolve the chambers. But Duverger's expression was accepted in the scientific community. In popular debates, however, 'presidial' or even 'presidential government' still occurs for this type of representative government. A survey has shown that even six percent of the German deputies wrongly classified the German parliamentary system.

4.3.3 Typologies in Comparative Social Sciences

The problem of 'miscomparing by misclassification' has been overcome by the construction of fourfold matrices which allow to put at least two classificatory elements (developed by Arend Lijphart) into relation to each other and to subsume various countries in four fields (cf. matrix)

Historical types			
Comte	Theological era	Metaphysical era	Positive era
Morgan	Wildness	Barbarian	Civilisation
Engels	Communist original society	Exploiting societies (slave holding, feudal, capitalist societies)	Communism
Theoretical types			
Weber	Traditional rule	Charismatic rule	Rational rule
Almond (styles of interest aggregation)	Absolute value orientation	Traditional style	Pragmatic bargaining style
Almond/Verba (political cultures)	Subject culture	Parochial culture	Participant culture
Duverger (party types)	Cell, militia	Traditional committee of notables	Mobilized party, organized in sections
Apter (types of developing regimes)	Mobilization regime	Consociational regime	Modernizing autocracy

4.3.4 Matrix: Institutional Mix for the Mitigation of Territorial Conflicts

The more complex such a matrix is developed, the more the comparing scholar has to be aware that such instant pictures can change quickly. British devolution, for instance, since the institutionalization of parliaments for Scotland and Wales have developed in the direction of Spain—with the exception of electoral law. If proportional electoral system—under the pressure of the European Union—one day is accepted even for British elections the whole type might shift in the above quarter of the matrix. Russia with its mixed type of an electoral law and with different levels of equality of the rights of the federal units is somewhere in between the four fields of the matrix.

Federalism			
		Equal rights	Unequal rights
Electoral law	Proportional	Moderately egalitarian Germany, Austria Egalitarian: Switzerland, Belgium	Unequal autonomies: Spain, Italy
		Russia	
	Majoritarian	Equal states' rights: USA	Devolution: Great Britain

4.3.5 From the Comparisons of 'Polities' and 'Politics' to Comparisons of 'Policies'

Comparative politics overcame the remainders of institutional typologies in the 1970s and 1980s when it turned from politics to the policies. Durkheim's study on *The Suicide* (1897) was a pioneering work in comparisons with aggregate data. Input–output comparisons were clearly inspired by the economic sciences. The main question was: "does politics matter?". Comparative politics influenced by political economy even suggested that the difference between capitalist and communist systems in the light of modernisation theory was exaggerated by the propaganda of both systems in a bipolar world of confrontation. Socio-economic determinism of Western theories of convergence of the systems since the 1970s had some traits in common with economy-centred Marxism which for reasons of methodology was not accepted.

By the early 1970s comparative public policy had emerged as a recognized sub-discipline within political science. The timing came not as a surprise. The world-wide intellectual unrest from Berkeley to Berlin had mobilized scholars to support an activist image of the state. Most of the scholars—except Huntington who wrote from a neo-conservative point of view—were liberal leftists. Though many of these comparativists had some 'Social-democratic bias' and admired the Scandinavian welfare state, they were tired of mere ideological discussions. The grand debates between Neo-Marxism and Critical Rationalism were abandoned. Scholars from different meta-theoretical schools agreed to discuss no longer abstract constructs but rather study the needs of groups in the society. "The state is better described by its policies than by its principles and alleged norms of individual choice and preference" (Ashford 1977: 572) was a widely shared assumption, especially in American political science, less so in Europe where the remainders of ideological debates were still stronger.

The behavioralist approach in the 1950s and 1960s frequently started in survey studies with individuals. Therefore, it was normally less open to transnational comparisons. The predicament of the small number of cases for comparison drove research into studies of cities or parts of the cities in the Californian Bay area or in New York. Przeworski/Teune (1970) pleaded for concentration on the subunits of a political system for comparison. But the result sometimes came close to a new parochialism which could be dubbed "The Westside story of comparative politics", because research concentrated on comparisons of hospitals or school districts in Westside New York. One further result of this kind of evolution was the abandoning of theories in comparative research. The new policy-orientation in the 1970s again concentrated on transnational comparisons. A model was created when Heidenheimer et al. (1975: V) in 1975 got an award for the best political science publication. The authors defined: "Comparative public policy is the cross-national study of how, why, and to what effect government policies are developed". The two schools which initiated comparative research continued to fight each other also in the subfield of 'comparative public policy'. The Heidenheimer-Heclo school was criticised for its descriptions without theories. The 'Quasi-Eastonians' in the school of Dye (1966) and Hofferbert on the other hand overcame the lack of theory by adopting abundantly the terminology of structural–functional system's theory. On the ground of a modernisation theory this approach had one assumption in common with the Neo-Marxists: levels of economic development were more important than the political characteristics of individual states. In the USA this assumption was mainly tested by comparisons among the policy performance of states. Wilensky (1975: XIII) upheld the hypothesis about the centrality of economic development for the provision of social services: "Economic growth and its demographic and bureaucratic outcomes are the root causes of the general emergence of the welfare state". Only European-born scholars, such as King (1973: 423), counter-veiled this widespread consensus on the priority of economic development by asserting that ideas constitute a sufficient condition for explaining the variance in policy performance. Most scholars combined the importance of ideas with the focus on elite groups or answered positively the question 'do parties matter?' The discovery of corporatism in the 1970s linked the elite approach with an emphasis on interest groups rather than on parties alone.

The Heidenheimer-Heclo school, based on historical and institutional studies, had the virtue of not neglecting 'politics' and the actors of decision, or reducing them to a kind of 'black box' for the production of a policy outcome which had little causal links to the decision-making process. Parties and interest groups were analyzed as important. Increasingly the role of the administrators was discovered who were the main actors in an intermediary stage of decision. They operated between *input* and *output* of the political system and their contribution has been called 'within-put'. The variance in the output of the compared systems was frequently explained in terms of rather vague special institutions, such as corporatism in Scandinavia or consensus democracy in Switzerland and the Benelux countries. Only the enlightened neo-institutionalism of the 1980s has stopped to look for unilateral causal relationships between two variables in the polity system and the policy output. But the inclination of comparative public policy for historical determination of policy outputs was mostly preserved.

Path dependence was discovered to explain why so many rising expectations for reform had failed in the 1970s. Restrictions, generated by historical developments and institutional barriers, left only 'narrow corridors' and *windows of opportunity* for policy action. To avoid a new kind of historicism it was important that comparative research was growing in order to avoid 'culture-bound generalizations' as a danger of one country studies. The hopes for reform proved to be dependent on the type of decision which has been classified. Different types of conflicts give rise to different types of legislative response and measures. 'Policy determines politics' was an exaggerated slogan by Lowi (1964) frequently tested in comparative politics. Lowi's typology initially showed a trinitarian design (regulative, distributive and redistributive). Later he added a fourth type: constituent policy. The elements of the typology were, however, not all on the same logical level. If we differentiate between regulative (restrictive limitation of rights, regulative laws, neutral to the question of gain and loss, extensive measures aiming at an enlargement of rights) and distributive levels of decision (protective, distributive, and redistributive measures) we end up with a six-fold typology (Beyme 1998: 5f). The typologies of policy fields and instruments of politics were soon connected by the network approach. A new slogan '*network determines policy*' was launched, but the differences between network theory and Lowi's assumptions were slight. Both predominantly saw a determination of the policy-output by interest groups and other actors in the 'cosy triangles' of the decision-making process (deputies, interest groups, and administrators who prepare implementation) even if they start from policies as an 'independent variable'.

4.4 Comparison in Post-modern Theories

Post-modernism in comparative political theory is not conceived as a completely new paradigm. Most reasonable post-modernists accept post-modernism only as a mere stage of modernity which implements its basic principles in a more consequential and systematic way than classical modernity. It cannot be equated with post-materialism or with certain processes of further differentiation and individualization which may lead to further decline of the old class social stratification and end up in theories of 'life style'. Post-modernity is a set of theoretical assumptions rather than a clearly discernible new structure of society. The hopes which were widespread in those European countries where ecological parties were strong, that a new 'society of movements' might develop, failed to materialize. The new social movements in most systems were strong in the phase of agenda setting. But decision and implementation was predominantly directed by the traditional organizations.

Post-modernism strengthened thinking in terms of constructivism. Durkheim's assumption that sociology per se is comparative was most eagerly adopted in post-modern autopoietic theories of systems. But Durkheim was still a realist and not yet a constructivist. 'Le fait social', the social fact, was his basic assumption but it in spite of his realistic way of thinking it was a kind of 'construct'. Post-modernist theories sometimes referred to Durkheim's approach to comparative science. The comparative method was no special approach for Luhmann, because he suspected that it aimed at a 'normative ontological framework'. He emphasized instead that comparative aspects had to be kept 'variable'. In autopoietic system's theory comparisons were not concerned with facts. In a society without a steering centre only the 'codes' which determined the development of the subsystems could be compared. But radically different codes (government and opposition in politics, true—not true in science, legal—illegal in law, beautiful—ugly in the arts) can hardly be compared since they function according to radically different logics. Systems and subsystems which evolve according to different codes can only 'observe' but not 'influence' each other. Adaptations from one system in another are hardly feasible. Thus, the main impetus for comparative politics was given up.

Post-modern theories, such as Foucault's (1969) *Archaeology of knowledge*, looked for variety. The 'summing up notions' in the 'archaeological comparisons' aimed at further pluralisation of discourses. The critical approach to comparative politics in post-modern thinking was overdue. But 'thinking in fragments' finally leads 'ad absurdum' since a controlled comparison is no longer feasible. In autopoietic theories of systems comparisons are close to 'pathology' in biology. The principle of degenerated cells are compared with sound elements of an organism.

4.5 Conclusions: Political Influences on the Evolution of Comparative Politics

The evolution of comparative politics is not—as it has sometimes been presented in the literature—a clear evolution of subsequent paradigms. There is permanent change in the perception of the needs of scientific comparisons, but hardly ever a dominant mainstream can be traced which deserves Kuhn's mostly overstretched term of a 'paradigm'. There are, however, phases in the relationship of political theory toward comparisons. The eras of 'pre-modernity', 'classical modernity' and 'post-modernism' show differences in the application of comparative methods. Pre-modern scholars mostly used comparison in an anecdotic way or deducted characteristics from human nature or certain forms of rule (e.g. the Roman Empire vs. Greek states). Only in the nineteenth century a historicist approach, believing in a teleological development of the political systems was spreading. Comparisons had to show the influence of a dominant factor such as demography or economy on the political systems. Comparison was predominantly applied to 'polities', rarely to 'politics', and rather seldom to the 'policies' in the respective systems.

The era of classical modernity for the first time developed rigorous criteria for scientific comparison, no longer confounding evolution and history, theory and practice and accepting that the political subsystem was no longer steering a whole society. Post-modern theories were still more strongly aiming at variety and doing away with the remainders of 'reification' of phenomena in classical modernity. 'Communication' as a key concept and the assumption of constructed mutual perceptions changed the mood of comparative scholars. The authors of classical modernity also for the first time tried to develop logically consistent typologies as a tool for developing hypotheses.

Two concepts dominated comparative research which can be traced back to Max Weber's historical institutional comparisons or to Émile Durkheim's early system's approach, starting from the assumption that there is no special field of comparative social science. Sociology or political science were considered as comparative *per se*. The second line of development—under the impact of economic theory—was more interested in forcasting future developments than in realistic description of facts.

Theories and methodological approaches do not arise out of a blue sky. Methods of comparative politics proved to be influenced by political events. After 1945 American interest in foreign countries called for a new interest in foreign institutions. Soon, however, the 'behavioral revolution' discriminated the old institutionalism in the tradition of Herman Finer or Carl J. Friedrich. The counter-movement emphasized the comparison of dynamics of politics and political behaviour. The emphasis shifted to interest groups and political movements. As a consequence of a world-wide unrest and disobedience among the youth in the late 1960s new critical-dialectical and sometimes Marxist theories challenged the alleged conservatism of the behavioralist mainstream which was blamed only duplicating the alienated political world by its surveys. Political science discovered that the institutions were maladapted but remained unchallenged by behavioralists and functionalists. The scientific revolt started under the label of 'critique of parliamentarianism' and soon ended up in a dogmatic new political economy of revolution.

When the revolt petered out *neo-institutionalism and policy analysis* became a minimal common denominator of leftists and main stream scholars. The decline of communist regimes facilitated the change in theories and methods of comparative politics. New democracies were founded. In the era of the behavioral revolution nobody would have dared to talk about 'constitutional engineering'. The breakdown of dictatorship since the 1970s and 1980s opened a need for discussing old and new institutions. Even 'rational choice institutionalism' was applied to the study of who took which option in a case of creating a new constitutional order. Some former leftist scholars, such as Jon Elster, Claus Offe or Adam Przeworski developed an approach, dubbed as 'rational choice Marxism' which no longer accepted a 'telos of history' but worked out alternative options of the new elites in a post-communist world.

The breakdown of communism re-encouraged most distant systems designs. Transitions to democracy were compared in different areas and in different times. The theories of modernization and transition to democracy were mostly modelled on developments in Southern Europe and South America in the 1970s. With the simultaneous transformations of both economic and political systems which had no precedent in recent history, old assumptions of modernization theories about the economic prerequisites of successful democratization are no longer applicable. New approaches were applied to the unique character of the peaceful revolutions in 1989, such as the testing of chaos theories from recent developments in biology and physics, although most of the applications remained metaphorical (Marks 1992).

Theories of various feasible roads to modernity in the tradition of Barrington Moore (1966) all of a sudden were outdated. On the other hand typologies of transitions, such as 'liberalization', 'democratization' and 'consolidation', were prematurely generalized. They were difficult to find in many areas of the world and proved to be a generalization of those former dictatorships in Southern Europe which consolidated quickly because they were soon integrated in the European Union (O'Donnell/Schmitter 1986). They might apply only to the Western tier of states in East-Central Europe which recently became members in the EU. 'Consolidology' as a subfield of 'transitology' became a new branch of comparative politics. The normative equation: "communism was a perversion of modernity, whereas post-communism is enlightened modernity" was quickly discovered as an untenable simplification. Soon comparativists discovered the 'defective democracies' which created another busy growth sector of comparativism (Merkel et al. 2003).

The acceptance of three worlds until 1989 included the acceptance of a plurality of systems. After the breakdown of Communism convergences are growing throughout the world. The concept of a 'world society' emphasized by post-modern variations of system's theory seems to be more plausible than before 1989. When the biologist Maturana in the early 1980s developed the hypothesis that no country in the world was able to develop fully its Communism because the world system was predominantly capitalist, this was taken as an abstract oddity of a non-social scientist Nevertheless the hypothesis proved to be true. Oddly enough theories of post-modernity which emphasize plurality face an increasing streamlining of the world.

The consequences of this development for the methodology of comparative politics are not yet fully recognized. Political science mostly ignored the 'grand debates' on the level of macro-systems. Business as usual continued. Political science stuck to the middle-level of mainly actor-oriented approaches. Recent revivals such as neo-institutionalism or rational choice had more impact on political science than the abstract peaks of a general theory in sociology. The status of institutionalism in comparative politics changed dramatically over the years. Institutionalism used to be an invective. At the turn of the century a new exaggeration was offered: "we are all institutionalists now" (Pierson/Skocpol 2002: 706).

Rational choice approaches initially served as a counter-balance against the dominance of behavioral studies in the 1960s. It was easily combined with a multi-level analysis and with an enlightened neo-institutionalism which spread in the 1980s. It was linked with new attempts to reach the scientific level of neighbouring social sciences, such as economics: "Rational choice institutionalism began as pure theft, lifting analytical tools from mathematics, operations research, and economics" (Shepsle 2006: 55). Rational choice approaches had the virtue to be applicable to any behavior from the most egoistic rationality to the most altruistic behavior of 'Saints'. Against the assumptions of macro theories about autopoietic systems the analysis of political actors remained meaningful. The strictly individualistic origin of the new approach was soon abandoned. Rational choice was applied also for collective entities and even for whole states. The research program of rational choice institutionalism conquered many departments in American universities, but never dominated in Europe. It was not unchallenged because of its abstractions, its simplifications, its analytical rigorism, neglecting context. 'Context' was a new catchword of comparative studies which turned back to individual cases or to comparisons in a middle-range historical perspective.

The 'evolution' of comparative politics was no self-steering development, but one more proved to be deeply influenced by political events. The predicament of political science was that its capacity to forecast major events was limited. There was hardly an anticipation of the student's rebellion, the 'third-worldism', the technological and ecological revolutions or the break-down of Communism. After 1945 the scope of American political science broadened to world-wide interests in area- and systems studies. The 'behavioral revolution' for a while seemed to develop into a kind of dominant 'paradigm'—but only in the United States. The events after 1968 challenged the naïve trust in democratic institutions and an 'enlightened neo-institutionalism' had its revival. In combination with rational choice approaches the two traditions merged in many ways. The crisis of policy-making under the impact of the oil crisis in 1973 strengthened the interest in transnational comparisons of public policies. The breakdown of communism renewed fields of comparisons which formerly were treated under the auspices of modernization theories with its simplified analysis of dominant factors. Democratization and consolidation of the new democracies was one major interest. But soon the failure of consolidation ended in a new boom of studying 'defective democracy' all over the world. If 'democracy' was the final normative target of comparative studies, the scientific community had to face the fact that there are many 'defective democracies' but nowhere a 'perfect democracy'—not even in the world of consolidated legal states and representative governments in North America and Western Europe.

References

Tocqueville, Alexis de, 1961: *De la démocratie en Amérique. Œuvres complètes*, Vol. 1 (Paris: Gallimard).

Almond, Gabriel A.; Verba, Sidney, 1963: *The Civic Culture: Political Attitudes and Democracy in Five Nations* (Princeton: Princeton University Press).

Amalrik, A. A., 1970: *Kann die Sowjetunion das Jahr 1984 erleben?* (Zürich: Diogenes).

Ashford, Douglas Journal, 1977: "Political Science and Policy Studies: Toward a Structural Solution", in: *Policy Studies*, 5: 570–583.

Bentley, Arthur F. 1949: *The Process of Government: A Study of Social Pressures* (Evanston, Ill.: Principia Press of Illinois).

Beyme, Klaus von, 1988: *Der Vergleich in der Politikwissenschaft* (Munich: Piper).

Beyme, Klaus von, 1998: *The Legislator: German Parliament as a Centre of Political Decision-Making* (Aldershot: Ashgate).

Boccalini, Traiano, 1614: *Ragguagli di parnaso. Centuria prima* (Milan: Battista Bidelli).

Botero, Giovanni, 1589: *Della ragion di stato* (Turin: UTET) (reprint 1948).

Chilcote, Ronald H., ^{2}1994: *Theories of Comparative Politics* (Boulder: Westview).

Daalder, Hans (Ed.), 1997: *Comparative European Politics. The Story of a Profession* (London: Pinter).

Downs, Anthony, 1957: *An Economic Theory of Democracy* (New York: Harper and Row).

Droysen, Johann Gustav, 1838: *Historik* (Darmstadt: Wissenschaftliche Buchgesellschaft) (reprint 1969).

Dye, Thomas R., 1966: *Politics, Economics, and the Public: Policy Outcomes in the American States* (Chicago: Rand McNally).

Easton, David, 1965: *A Framework for Political Analysis* (Englewood Cliffs: Prentice-Hall).

Durkheim, Émile, ^{11}1950: *Les règles de la méthode sociologique* (Paris: PUF).

Falter, Jürgen; Klingemann, Hans-Dieter, 1998: "Die deutsche Politikwissenschaft im Urteil der Fachvertreter", in: Greven, Michael Th. (Ed.): *Demokratie eine Kultur des Westens?* (Wiesbaden: Westdeutscher Verlag): 306–341.

Foucault, Michel, 1969: *L'archéologie du savoir* (Paris: Gallimard).

Friedrich, Carl J., 1963: *Man and His Government* (New York: McGraw-Hill).

Hancock, M. Donald, 1983: "Comparative Public Policy", in: Finifter, Ada (Ed.): *Political Science. The State of the Discipline* (Washington, D.C.: American Political Science Association): 283–308.

Heidenheimer, Arnold J.; Heclo, Hugh; Adams, Carolyn T., 1975: *Comparative Public Policy: The Politics of Social Choice in Europe and America* (New York: St. Martin's Press).

Keman, Hans (Ed.), 1993: *Comparative Politics: New Directions in Theory and Method* (Amsterdam: VU University Press).

King, Anthony, 1973: "Ideas, Institutions, and the Policies of Government: A Comparative Analysis", in: *British Journal of Political Science*, 3: 291–313, 409–423.

Küchenhoff, Erich, 1967: *Möglichkeiten und Grenzen begrifflicher Klarheit in der Staatsformenlehre* (Berlin: Duncker & Humblot).
Lowi, Theodore, 1964: "American Business, Public Policy, Case Studies, and Political Theory", in: *World Politics*, XVI: 677–715.
Luhmann, Niklas, 1970: *Soziologische Aufklärung* (Cologne: Westdeutscher Verlag).
Mair, Peter, 1996: "Comparative Politics: An Overview", in: Goodin, Robert E.; Klingemann, Hans-Dieter (Eds.): *A New Handbook of Political Science* (Oxford: Oxford University Press): 309–335.
Marks, Garry, 1992: "Rational Sources of Chaos in Democratic Transition", in: *American Behavioral Scientist*, 35: 397–421.
Merkel, Wolfgang et al., 2003: *Defekte Demokratie*, Vol. 1. Theorie (Opladen: Leske & Budrich).
Mill, John Stuart, 1843: *A System of Logic* (London: Longman) (reprint 1959).
Mill, John Stuart, M., 1859: "de Tocqueville and Democracy in America", Vol. 2 . (Dissertations and Discussions, London, John W. Parker).
Moore, Barrington, 1966: *Social Origins of Dictatorship and Democracy* (Boston: Beacon Press).
O'Donnell, Guillermo; Schmitter, Philippe (Eds.), 1986: *Transitions from Authoritarian Rule* (Baltimore: Johns Hopkins University Press).
Parsons, Talcott 1937, ^{2}1961: *The Structure of Social Action* (Glencoe, Ill.: Free Press).
Pierson, P.; Skocpol, Theda, 2002: "Historical Institutionalism in Contemporary Political Science", in: Katznelson. I.; Miller, H. V. (Eds.): *Political Science: State of the Discipline* (New York: Norton): 693–721.
Przeworski, Adam; Teune H.,1970: *The Logic of Comparative Inquiry* (New York: Wiley).
Ragin, Ch. C., 1987, 1989: *The Comparative Method: Moving Beyond Qualitative and Quantitative Strategies* (Berkeley: University of California Press).
Sartori, Giovanni, 1991: "Comparing and Miscomparing", in: *Journal of Theoretical Politics*, 3: 243–257.
Shepsle, Kenneth A, 2006: "Rational Choice Institutionalism", in: Rhodes, R. A. W. et al. (Eds.): *Political Institutions* (Oxford: Oxford University Press): 23–38.
Somit, Albert; Tanenhaus, Albert, 1964: *American Political Science. A Profile of a Discipline* (New York: Atherton).
Troeltsch, Ernst, 1922: *Der Historismus und seine Probleme* (Aalen: Scientia) (reprint 1961).
Wilensky, Harold, 1975: *The Welfare State and Equality: Structural and Ideological Roots of Public Expenditures* (Berkeley: University of California Press).

Chapter 5
Neglected Functions of Normative Political Theory in Empirical Research

5.1 Political Reasons for the Decline and Resurrection of Normative Political Theories

The theory of politics began as a part of ethics. Plato dealt with politics and ethics in the same book. Aristotle began to separate the two fields but nevertheless close links remained: *ethics* was the more *static*, *politics* the *dynamic* side of the same coin. On the continent, political theories after 1945 were frequently subsumed under three labels: the empirical-analytical school, the dialectical school, and the neo-Aristotelian school. The last two labels pointed to predominantly normative theories. Nevertheless, normative theory was an outsider in the grand debates. No other century has contributed so little to questions of ethics and politics, whether Husserl or Heidegger, Wittgenstein or Whitehead, Bergson or Pierce or the critical rationalism of Karl Popper is debated (Ballestrem et al. 1990: 8).

The rather militant anti-normativism in the social sciences has been explained by scientific and political reasons. A scientific reason was the *professionalization* of political science, which had developed from various origins in neighbouring disciplines. Behaviouralism abandoned holistic notions such as 'the state' and turned to small units such as individuals and groups. *Holism* was considered as an 'unworkable doctrine' (Philipps 1976: 123). Even *functionalism* with its implications of stability and balance was considered as a normative doctrine close to *positivistic organicism* and was not welcomed as a contribution to the analytical school of thinking by many rigorous behaviouralists.[1]

Only on the continent did the experience of totalitarianism after 1945 lead to a short revival of normative thinking. As far as the United States contributed to this debate, mostly immigrants from Europe, such as Hannah Arendt, Leo Strauss or Voegelin were involved. Analytical thinkers in the United States mostly did not believe that normative political ethics should be the appropriate answer to

[1] This text was first presented at the University of Vienna as "Revival of normative theory in empirical research" and has not previously been published.

K. von Beyme, *Klaus von Beyme*, SpringerBriefs on Pioneers in Science and Practice 14,
DOI: 10.1007/978-3-319-01535-4_5, © The Author(s) 2014

totalitarianism because the close links between morals and politics was for them a driving force of the development towards totalitarianism in the 1930s and 1940s (Moore 1978: 399). *Differentiation* became a basic concept of political theory—not *holism*.

Conservative normative theories of politics normally opted for a *mixed government* with strong emphasis on the *Rechtsstaat* (legal state) rather than on democratic participation. The American *political revolution* was the model—not the *social revolution* intended by revolutionaries in France since 1789 and in Russia since 1917. *Radical* normative theories, on the other hand, criticized precisely this USA-centered image of a 'good polity'.

The critical-dialectical school, particularly strong in post-war Germany, was normative without offering a model, and concentrated on developing a political theory which was not the mere theoretical duplication of a miserable social reality. Communism as a model since Marcuse was no longer feasible because its rationalistic and technocratic bias suffered from the same deficiencies as the 'imperialist democracy' in the West. Critics complained that the bipolar constellation in the world created on the one hand a hidden collusion of theoretical principles in the first and second worlds and on the other a political divergence of two competing ideologies which developed anti-communism in the West into a general anti-normative thinking (Rödel et al. 1989: 11f).

The erosion of communist systems moved normative models back into the centre. Concepts such as 'virtue of citizens' and 'civil society' proved to be not just utopian dreams; they had contributed to the peaceful revolutions in Eastern Europe and the victory of democracy over a bureaucratic and sclerotic communist system (Cohen et al. 1992: 15). Long before the collapse of the last European dictatorships, a change of paradigm had developed since Rawls's *Theory of Justice*, which was powerful and convincing—though rather unpolitical (Ricci 1984: 321f). The new normative debate from the early 1980s, as an answer by those thinkers labelled 'communitarians' to the liberal formalism and proceduralism of the Rawlsians, brought *politics* back into a discussion which had started as a debate on the fundamentals of the *polity*. The remarkable side of this new wave of political philosophy was that no monopoly of justification and certainty was erected. The debate was really interdisciplinary and some of the leading figures, such as Walzer or Sandel, held chairs of political science rather than philosophy. Political science even in America has been a hidden resort of normative thinking at a time of the triumph of scientistic concepts. Even the strongholds of the behavioural revolt, the political science departments of Ann Arbor, Chicago or other schools, have permanently included about 10 % of the chairs of political philosophy and normative reasoning in the light of a history of political ideas (figures in: Beyme 1988). The flexibility of political science departments offered some of the communitarian philosophers a wider audience than they would have had in highly specialized departments of philosophy.

Some conservative torchbearers among the communitarians who criticized the dominant liberal theories, such as MacIntyre (1981), directly referred to theories of virtue which were widely discussed among the neo-Aristotelian schools in

Europe after the war. *Virtue of citizens* was developed into a counter-concept to the liberal concept of *justice*. The protagonists of the two positions were streamlined into radical alternative positions: atomist abstract and universalist liberalism versus humanist communitarianism starting with the concrete situations of life and reflecting the needs of divergent political actors (O'Neill 1996). *Postmodern thinking in fragments*, which originally had nothing to do with the new search for justice, provided intellectual leverage for communitarian ideas. Moreover, the debates about *political correctness* attracted political theories which emphasized particularist rights against universalist generalizations. The *new republicanism* contradicted the liberal assumption made since Kant that *material justice*—not only formal procedural justice—is superior to ideas about the common good. Politics of justice was more than the utilitarian sum of all individual preferences. Freedom and participation were intimately linked in these new theories, whereas liberalism considered freedom as the higher value and thought of participation only as an accidental instrument of the implementation of freedom. Most communitarians emphasized self-government as a condition of liberty. The radical alternative of virtue versus justice in the long run was overcome: *virtue with justice* became a kind of new common denominator.

The theories of civil society and citizenship ended the ridicule of concepts such as virtue by the analytical and empirical scholars. Habermas (1981, vol. 1: 18) in his survey on the shortcomings of modern social theories blamed political science for having abandoned the normatism of natural law and for excluding questions of legitimacy. Legitimacy since Max Weber had been reduced to a 'belief in legitimation' which could be studied with the help of empirical survey methods. In the same year that Habermas wrote this harsh criticism of a whole discipline, American social theory accomplished a change of paradigm. Liberalism and communitarianism continued to criticize each other, but they were basically at one in renouncing the empirical *utilitarian* model of democracy. Both demanded a theory of democratic legitimacy and justice. The new normativism had a broad and pluralistic consent. Normativists were reborn, as Cohen and Arato said in their seminal survey of civil society theories.

5.2 The 'Rapprochement' Between Normative Theory and Empirical Research

In the history of political theories a normative element has never been completely abandoned. Sabine defined political theory in terms of statements about facts, statements about causal relations, and prognoses about likely developments in the future. The third element was reflection on what 'ought to be'. Sabine's assumptions were based on the pragmatic tradition from Dewey to Mead. Only dogmatic rational choice thinkers thought operation no. 2, prognoses about forthcoming developments and the accuracy of their predictions, more important than the reality of their assumptions (Downs 1957: 21). The most anti-normative approach, the neo-Kantian division between 'what is' and 'what ought to be' did not exclude normative judgements, but it did not consider them as part of the scientific work *stricto* sensu. The American pragmatic tradition predominantly did not accept this division of what there *is* and what there *ought to be*. Even early behaviouralism was close enough to the progressive movement in America to include normative postulates in its intellectual operations. Critical policy analysis and the dissenters to the mainstream of the 1960s organized themselves into the 'Caucus' within the American Political Science Association, and started from normative postulates even before analysing political reality. Critical rationalists of Karl Popper's school on the other hand were convinced that normative reasoning is to be excluded from the central operations of analysis, and is only admissible at the stage of hypothesis formation and of drawing conclusions from empirical analysis. With increasing numbers of defectors from neopositivism and Popper's rigorous orthodoxy of falsification, the old Weberian postulate for *Wertfreiheit* (value-free science) was reduced to historical relativism: value-free science as a postulate made sense only at a certain historical stage of intellectual development. The postulate of value-free science served as an instrument for professionalizing young scientists and for protecting science against interference from the state. It also served as a buffer against the 'false friends' among those scholars sharing a certain political position who tried to transform their 'friends' into zealots of a 'common cause'. Value-free science was a kind of armistice between different schools of thought in the age of ideologies and between scientists and politicians (Beck 1974: 208ff). Value-free science was a programme of action, but the scientists knew that it could only partially be implemented. Theoretical concept formation in many cases meant normative thinking, especially when holistic notions of integrating a fragmented society were involved, such as 'democracy', 'civic culture' or 'civil society'. Normative 'ascetism' did not prevent the inclusion of normative elements in the analysis because even so-called social reality was "impregnated by normative elements" (Peters 1993: 24f).

The rapprochement of normative political philosophy and empirical political theory was facilitated by the decline of neopositivism—which had served as a rampart against the historicist normativism of the dialectical school and the neo-Marxists. It was no longer urgently needed when Marxism eroded a decade before Communist systems withered away. Postmodernist thinking in fragments was hostile to the holistic assumptions of Weber's value-free science and rationalism. Moreover, the new theoretical trends were moving from *realism* and the idea that the concepts corresponded with social reality to various shades of *constructivist thinking*. The moderate variation of constructivism does not assume that all concepts are conventional agreements among scholars and that there is no social reality. They rather knew that crude empiricism was wrong and that empirical assumptions and a kind of 'fictitious reality' are constructed. The school of rational choice did not accept that 'motivating political morality' was behind social reality and that therefore theoretical sentences cannot be formulated unless time and context is considered (Goodin 1992: 150). But in a parallel way certain mathematically oriented schools of thought in social science developed the idea that there is no social reality. Rational choice scholars construct a social reality as an approach to the world as it *may be* and as it can be predicted in its development. Like normative thinkers, they avoid contact with descriptive reality. Normative assumptions are fully included in this constructed image of the world: the goodness of Mother Teresa can be maximized according to the same rules as the most reckless 'casino capitalism' of mafia entrepreneurs.

Normative assumptions in many approaches were the starting point for any empirical analysis. The formation of concepts for analysis was discovered to be not very far from normative deduction. Mutual tolerance between political philosophy and empirical political theory was growing. Policy analysis—starting with desired goals in development—has contributed considerably to this rapprochement.

Normative thinkers have complained for decades that empirical research has marginalized them and kept them in the ghetto of 'powerful book learning' with no relevance to empirical science. After the *rapprochement* their influence on political science has grown. Nevertheless, they were not satisfied. As with theology in some progressive divinity schools—apart from the hard core of dogmatic theory—the normativists turned to a kind of *soft social science,* reflecting everything under the sun in terms of the social sciences, only without the rigorous methods applied by them. They frequently turned to semi-empirical work, digging into many details of empirical analysis. This was very obvious in the case of the communitarians who had started with a criticism of the abstract and formalistic notions of Rawls's theory of justice and tried to be closer to the political reality of their respective countries.

The revival of normative political theory was accompanied by a paradoxical development: the more open empirical research, especially in policy analysis, was to normative theory, the less normative theory political philosophy dared to offer. Habermas complained that the 'idealistic content' of normative theory 'melted under the sun of experiences in social science'. Habermas proved to be a good example for the development he analysed. For decades, his theory of discourse was denounced as an idealistic utopia far from the experiences of empirical communication theory. Now he denounced the communitarians as being 'too idealistic'—even when they pretended to work on the basis of Habermas' theory of discourse. The new political ethics among the communitarians was doomed to failure according to Habermas because there is no agreement on one system of political ethics. The plurality of positions needs mediation between interests, which cannot be achieved via ethical discourses. This criticism of communitarianism sounded just like earlier criticisms of Habermas' theories by empirical scholars. The argument that his image of man in society was too optimistic, Habermas passed on to the communitarians. Hume's device 'rules not men'—a classical conservative and liberal argument against the preachers of virtue—was now directed against the new theory of virtue, since Habermas rediscovered the *Rechtsstaat*, the constitutional legal state. Thus, he came close to the slogan of 'freedom instead of virtue' which was developed by the older liberals in opposition to the post-war normativists.

There was a second paradox in the development of political theory and political philosophy: Utopias became acceptable since the Communist systems—originally built on utopian political thought—perished and no longer discredited them. Deideologization in normative theory-building even made analytical thinkers more tolerant towards normativism. Empiricists started encouraging the normative school with paternalistic encouragements: 'courage to develop utopia' because they knew that utopias no longer endangered the status quo of society.

Mannheim (1952: 225) had tried to declare the *decline of ideology* as a positive movement. The *decline of utopian thinking*, however, was a vision of horror for Mannheim: "The perishing of utopia created a kind of static rationalism in which man is degraded to an object". The negative notion of ideology—opposed to the positive notion of a progressive *Weltanschauung* in Marxist thought—was behind this misleading dichotomy. Utopia was the necessary indispensable normative dream of a society grounded on justice. Ideology, on the other hand, was only the appearance of justice, used to hide the partial interests of one social group. This dichotomy is hardly tenable. Ideologies are more partial normative systems, but they are less easily recognized than utopias. Communism in the phase of erosion has demonstrated the dangers of ideology. The utopian dream of a Communist society was first postponed and later abandoned. An ideology of *real socialism* was opposed to Maoist and Castrist utopias which cannot be implemented. *Mature socialism* as an ideology replaced the utopia of Communist society and attracted more acceptance from the non-Communists than the older utopia which had fascinated many intellectuals in the West but soon was discovered as the 'God that failed'.

Bell (1966: 405) was another thinker in the tradition of the old left who hailed utopias and discriminated between ideologies. 'The end of ideology' for him amounted to a still greater need for utopia:

> The end of ideology is not—should not be—the end of utopia as well. If anything, one can begin anew the discussion of utopia only being aware of the trap of ideology. The point is that ideologists are 'terrible simplifiers'... There is now, more than ever, some need for utopia, in the sense that men need—as they have always needed—some intelligence. Yet the ladder to the City of Heaven can no longer be a 'faith ladder', but an empirical one: a utopia has to specify where one wants to go, how to get there, the costs of the enterprise and some realization of, and justification for the determination of who is to pay.

Again the positive assessment of utopia and the negative evaluation of ideology is not very convincing. Utopias in the classical sense never contained empirical elements to indicate 'how to get there' and the 'costs of the enterprise'. Normally ideologies have added this element of strategic thinking to the utopian core beliefs—and this has discredited the normative theory in the long run because the strategical and tactical elements tended to grow at the expense of the normative core of the utopia. The empirical test of utopia demanded by Daniel Bell will probably lead to discarding the utopia or to ideological simplification in order to show that parts of the utopia can be implemented without exaggerated costs.

In the new 'rapprochement' of empirical and normative thought, utopias became more acceptable again. But their normative content was mostly exaggerated, unless it be that some kind of 'State of God' was called utopia. The classical utopias of early modern times from Thomas More down to Harrington contain a lot of empirical elements together with some normative postulates. The 'language of Aesop' in times of absolutist rule forced thinkers to hide empirical analysis behind the exploration of some distant utopian land which quite frequently had visible traits of the country for which the work was written. Thoughts that might have been perceived as utopia at their time, such as More's hopes for a humane treatment of criminals or Harrington's proposals for reorganizing agricultural property for a gentry which supported the system, became more empirical after a century than they were at the time of writing. Other elements of utopias, on the other hand, were regressive concepts in a society which declined over the course of the century, like many utopias in the late Middle Ages which hoped to revive the old system of a functioning society of estates, each respecting the rights of the other strata and powers. At the end of such a development, fictitious travel reports were written, such as Montesquieu's 'Lettres persanes', which hardly contained utopian elements and were just contemporary thoughts on negative developments disguised under exotic references to distant countries in order to mislead censorship.

This erosion of utopia since the period of renaissance was accelerated after the end of totalitarian ideologies. *Autopoiesis*—a postmodern version of systems theory after Parsons—was probably the most 'defeatist' theory concerning normative thought. Normative postulates were frequently ridiculed by Luhmann as 'remnants of old Europe'. Like planning and forecasting, the development of ideologies was declared to be futile. Observation of the evolution was the only operation of political theory that made sense. Habermas has criticized this cynical postmodernism which amounted to a new decisionist competence of experts in what Luhmann called 'administration'—oddly enough comprising politics. But this criticism proved to be a rearguard fight: the normative content of the theory of discourse was shrinking from book to book as Habermas wrote. The boom in ethics remained largely theoretical. Ethics commissions were established when new problems arose where finding a consensual decision seemed to be difficult. The new moral thrust was even a kind of compensation of the fact that a fragmented society can no longer agree on a rational strategy for solving a problem. The normative results of these strategies of evasion are, however, as fragmented as the society which tries to delegate moral responsibility: ethical commissions create fragmented special ethics for every field of activity or scientific discipline. Luhmann's cynical theory explains the functions of normative reasoning as a compensation for the lack of an unified ethical system. Morality in a fragmented society seems to have the necessary bipolar code which establishes a distinct subsystem: good or bad. In spite of these simple criteria, morals do not constitute an independent subsystem in modern society (Luhmann 1993: 33). This weakness is compensated for and morals serve as a kind of reinforcement for the values in other subsystems: political morality is invoked in cases of corruption and scandals in politics and contributes to a self-purification of the political system. The subsystem of politics cannot be transformed into a moral entity. Morals serve only to preserve the code of functioning of the subsystem by sacrificing some individuals who abuse the possibilities of politics.

Normative theory in such a concept of social theory, which starts from the assumption of further differentiation and fragmentation, parallel to the evolution of the whole world where the elements are drifting apart towards entrophy, only has the function of reinforcing the ethical code of the system which is the least common denominator. Each normative postulate provokes counter-postulates. The normative theory of politics thus has the function of finding those minimal agreements acceptable to most members of society or of one of its subsystems. Normative thought in this perception has a peacekeeping function in a situation which is close to a civil war of competing norms. This contributes, however, to a development in which the utopian content of normative thought is further shrinking and political ethics is oriented towards consensus and political reality.

The new normative political philosophy does not require a belief in metaphysics. Voegelin (1965: 14), torchbearer of the old normativists, who founded practically a sect of true believers, still demanded belief in an ontology of strata that included the divine sphere, to make it relevant for the lower spheres of society within the world. It was not by chance that most empirical political scientists who tried to read such postulates remained 'pre-clear' and outside the in-group, and (according to the master's view) among the mass of people who had only *doxai* (opinions) to offer instead of insight and wisdom. But there were softer forms of normativism—from Hannah Arendt and Michael Oakeshott to the Freiburg school in the tradition of neo-Aristotelian 'practical philosophy'. Some of them have been revived by the waves of neo-normative thinking and integrated into empirical political science. The old controversy between *episteme* (science) and the search for *phronesis*, political and social prudence, have fortunately not been revived. The contemporary consensus is rather: both episteme and phronesis!

5.3 The Normative Duplication of Social and Political Reality

Critical dialectical thinkers such as Habermas have been 'Weberized'. Dogmatic American liberalism has been communitarized (Bell 1993: 8). Rawls made some concessions to the communitarians. Walzer has accepted that communitarianism is not a fundamental opposition to liberalism but only a corrective in times when liberalism is uncritically exaggerated. Communitarianism serves to strengthen the healing forces within liberalism in itself.

Normative thinking in political theory withdrew into a kind of 'constitutional engineering' after complaints about *ungovernability* had been abandoned by the conservative camp. Governing has changed to *networking*—but this is precisely the reason why non-governability can no longer be imputed. Right-wing technocrats do not offer far-reaching political vistas. The debate is restricted to a few corrections of the existing systems such as popular election of the president and a switch to a semi-presidential system, combined with a more majoritarian electoral law. Some countries in crisis have half-heartedly introduced some changes towards these normative postulates: Israel created the worst of two worlds with the popular election of the prime minister and Italy changed its electoral law with a kind of *reservatio mentalis* which has been dubbed: 'maggioritario ma non troppo' (Bartolini et al. 1995). Conservative and radical reformers sometimes agree—except in Germany which stubbornly and wrongly thinks that the Weimar Republic disintegrated because of popular referenda—that some plebiscitarian elements should be introduced in order to overcome the alienation of the citizens from their political system, the so-called *Politikverdrossenheit*. The normative image of a strong democracy developed by Barber (1984) was widely discussed but in tune with the trends of normative thinking was rather 'homeopathic'. Theory according to Barber should be coherent but realistic. Existing representative institutions should be revitalized by new participatory possibilities for the citizens. The de-professionalization of politics was demanded in many variations of homoeopathic political theory from Barber to Beck. This kind of reformist pragmatism is reasonable—but hardly motivating for the alienated citizens, who escape into subsystems of privacy. Former radicals among the dialectical school of thought, such as Offe, restrict the reforms by what he calls the 'auto-paternalistic mental reservation'. Reformers need to renounce means of participation which confuse the normal representative process of decision-making by imposing any topic at any time if minorities so desire.

There are still some models of participatory democracy being discussed, but most of them have abandoned the assumption that participation is the highest value per se because it creates an 'outside interest' for the citizens and transcends the concept of merely private life (Naschold 1969: 51). Mostly, realistic political theory has accepted that politics is a rather marginal field of interest for most people. We have to explain why some people are eagerly interested in politics rather than why most people have more private interests. This does not mean that political theory has to fall back to the opposite extreme and impute with Harold Lasswell a pathological background to those who demonstrate political ambition. Utilitarian calculations are admitted in postmodern normative political theory and no longer 'denigrate' the values of participatory democracy. Most people neither have the resources nor do they see the necessary rewards for participating in politics in order to attain more than a short-time goal. Since the rise of the new social movements, *ad hoc politics*, mobilization for a limited purpose and outside the big machines of parties or organized pressure groups, is not only accepted but even frequently considered as the most appropriate method of democratic participation. Not *maximizing* but rather *optimizing* democratic participation is the goal of normative democratic theory (Schmalz-Bruns 1995: 17).

Democratic theory has recently been characterized by a certain 'normative defeatism'. The meta-theories behind this development have mostly abandoned the idea that *truth* can be attained by reasoning and are satisfied with a conventional idea of truth: truth is what most people can agree on. With the reduction of truth to convention, the normative thrust of theory is considerably reduced. This can be shown in the theoretical evaluation of politics by committees of experts as well as by ethical commissions: since agreement about the scientific feasibility of a political measure is not easy to attain, the *acceptance* of politics by the people is increasingly the supreme criterion for evaluation. Lincoln's Gettysburg formula still included three elements of democracy on an equal basis: politics *of* the people, *by* the people and *for* the people. With shrinking political participation the value 'for' the people and the *responsiveness* of politicians is the main indicator for good politics.

After this development, fundamental criticism of the 'system' is no longer feasible. The normative goals of democracy try to optimize democratic participation in the subsystems. 'Bargaining democracy', 'cooperation of networks', 'politization of subsystems' are the main concepts. '*Reflexive democracy*' is the new label—but its torchbearers no longer claim to protect the normative goals of their concept against the insights of empirical research (Schmalz-Bruns 1995: 153). It is not so much empirical theory which limits the claims of normative political philosophy. It is rather the self-restraint of normative theory itself which spreads after the decline of the Popper orthodoxy of critical rationalism. Habermas is the best-known model for a development which seemed to force a normative thinker to accumulate more and more empirical knowledge. The problem is, however, that these empirical findings are used in a highly selective way. Habermas' normative idea of discourse did not need empirical research into the *Lebenswelt* of small groups and families. The problem is rather that Habermas discusses the problem of 'life world' on the epistemological level of Husserl's phenomenology, on the level of social philosophy in the tradition of Schütz, and on the third level of empirical sociology. Only the top level was necessary for the normative derivation of his conclusions (Apel 1989: 22ff).

The obsession of normative thinkers from Habermas to Walzer with increasing the empirical plausibility of their concepts by heaping up analytical literature normally does not make the concepts used clearer. There is an infinite regression to auxiliary theories down to the sphere of linguistics, psychology of perception, or evolution of childhood. All this would have been unnecessary if Habermas had been satisfied that he developed—next to Rawls's theory of justice—the most encompassing normative theory of social life. Habermas seemed to be offended when he was said to have developed a normative political theory. As a matter of fact, his theories are much less acceptable in their empirical parts than in the normative core. Against the intentions of the author, the main influence of his theory came from the normative parts of his works.

There is no harm done if an eminent author writes two volumes instead of one to increase the empirical evidence and plausibility of his assumptions. The reader can overlook these parts and dig into the essence of a book. The negative impact of this kind of theoretical development is, however, that normative theory gets more and more affirmative the stronger it works with empirical facts. Again Habermas is a good example. His work was hailed as a forerunner of normative justification of the new social movements. Habermas was sceptical from the outset. But from the early 1990s he stressed that parliament should be the centre of all the political discourses and deliberative politics he advocated (Habermas 1992: 215). The thrust of civil society and its torchbearer, the social movement, is reduced to 'loopholes of reflection' and 'pluralization of group formation'. Empirical network analysis comes to the same conclusion—but without normative theories. *Reinvention of reality* is the result—not normative political theory (Cohen et al. 1994: 149ff). Beck (1993: 209) is a master of reinvention when he tries to 'invent the political', which means a new politics that not only applies rules but aims at a 'reflexive' policy that changes the rules. But his politization is rather tautological. Strikes and electoral politics are said to be old-fashioned. But what is new? Blockades and stoppages as his new instruments of reflexive democracy are, however, just called minor instruments which have always been used by the traditional social movements. *Network* is the new concept, which creates a minimal consensus between normative theorists and empirical analysts in political science.

The positive side of this development is that the social movements are no longer said to create 'authentic participation' and democracy. The new minimalism of participatory democracy aims at a competition of the modes of participation offered by the state and those offered by society (Held 1989: 182). The 'double strategy' of the traditional Leninists is deprived of its former revolutionary ambiguity. Only a peaceful competition of modes of participation is envisaged by the new models of participatory democracy. The new theories of democracy no longer accept Berelson's theory of 'stabilizing apathy', but rather develop models of moderate participation. Most of them know that there are empirical limits to participation because in certain areas theoretical fashions mobilize people and lead to "an overcrowding of a policy arena" (Schmalz-Bruns 1995: 250). Even Beck—so fond of 'subpolitization'—recognized that too much subpolitization can strengthen social confusion and lead to general powerlessness (Beck 1993: 233).

New theories about elites and political class, however, promote little confidence even in these limited forms of democratic participation. *Responsiveness of the elites* seems to be the only way of preventing the development of further political alienation and dissatisfaction (*Politikverdrossenheit*). Political philosophy thus came so close to developments in the political world that the political philosopher was in danger of abandoning philosophy and becoming an imitation of the politician in the realm of discourse. This tendency was obvious even in Rawls's later specifications of his theory—though there is also the opposite criticism that Rawls remained in the pure realm of "what ought to be" (Gerecke 1995: 44).

There is a final paradox developing: no Marxian verdict against images of the good society—sometimes compared with the iconoclastic mood of Jewish religion—is visible against the utopian socialists of our time. No neo-Kantian rigourism can uphold the eternal gap between what there *is* and what *ought to be*, which has prevented the development of normative political theory for almost a century—at least in Central Europe. Nevertheless, normative theory is mutilated by coming too close to reality. Walzer's 'spheres of justice' escaped this dilemma in a particular way (for an alternative view: Cohen 1986), but Etzioni's book has caused the normative ideas to deteriorate by digging into the details of 'neighbourhood watch groups' and a 'citizen-oriented police'. He even overlooks the dangers of this kind of 'concretism' and does not recognize that the social control of mobilized social groups can be more difficult than state control, which is mostly incomplete and inefficient. Political theory in the age of postmodernism has developed in a way which was criticized in the 1960s by the radical students' movement: political theory tends to become the mere duplication of reality—without creative normative fantasy. Science tends to follow the arts: modern or postmodern art no longer overcomes an unpleasant reality but from pop art to kitchen-sink art reproduces dreadful reality in a fragmented way.

5.4 Conclusion

Normative political theory is no longer completely neglected in political science as was the case in the period of the 'behavioural revolution'. Neo-contractualism was the great thrust of revival for normative theory. The concept of civil society made normativist theories reasonable. To support certain normative positions, ethnic or etatist theories are no longer necessary. Postmodern thinking in fragments swept away the hierarchical ontologies which characterized the neo-Aristotelian and neo-Thomistic normative theories after the Second World War.

Neo-contractualism abandoned thinking in terms of collective entities such as classes or nations. The individual returned to the centre of political theory. The mutual contractual relationship of individuals, recognizing their rights and duties, no longer needed metaphysical assumptions. Since Kant, the social contract has anyway only been a construct of reason. Mutual recognition originally seemed to restrict itself to a procedural ideal in John Rawls's terms. Communitarianism has added an element of material justice. Rational choice theories help to reconstruct the strategies of individuals, which are no longer restricted to maximizing personal benefits or even profits.

This extremely positive development was accompanied by the drawback that normative theories came too close to empirical findings. Analytic-empirical theories and normative theories tend to merge. If normative theory comes too close to empirical reality the result may be purely *democratic decisionism*. The former function of normative theory was to transcend the narrow framework of rationalist modernity.

Communism perished because the gap between the normative ideal of origins and 'goulash or Polski-Fiat communism' in reality was too big and the credibility of the systems was undermined. Democratic market societies have less ambitious positive normative ideas about a society of justice. But the negative freedom from state supervision, combined with options for individual social development, are normally respected. This does not mean that there are no gaps. In the United States this has been called the IvI gap (ideas versus institutions) (Huntington 1981: 39, 45). In waves, the normative basis of the American creed is violated and the institutions have to be purged in order to minimize the gap between ideals and institutions. For a long time, European politics had no basic consent underlying the institutions, and fundamental criticism and new normative theories had to aim at 'another Republic'. In the postmodern period, with the erosion of ideologies and the huge machines of membership parties in the old and the new democracies, Europe is getting closer to the pragmatic way in which American *Republicanism* has handled the normative problem. The rise of a non-ethnic consensus of some kind of *constitutional patriotism* is the minimal agreement of citizens on a normative ideal. It leaves room for normative theory—but excludes ideological alternatives and limits its functions to working towards overcoming the periodically rising gap between normative ideals and political reality.

References

Apel, Karl-Otto, 1989: "Normative Begründung der 'Kritischen Theorie durch Rekurs auf lebensweltliche Sittlichkeit?, in: Honneth, Axel et al. (Eds.): *Zwischenbetrachtungen im Prozeß der Aufklärung. Jürgen Habermas zum 60. Geburtstag* (Frankfurt: Suhrkamp): 15–65.

Ballestrem, Karl Graf; Ottmann, Henning (Eds.), 1990: *Politische Philosophie des 20. Jahrhunderts* (Munich: Oldenbourg).

Barber, Benjamin, 1984: *Strong Democracy* (Berkeley: University of California Press).

Bartolini, Stefano; D'Alimonte, Roberto (Eds.), 1995: *Maggioritario ma non Troppo* (Bologna: Il Mulino).

Beck, Ulrich, 1974: *Objektivität und Normativität. Die Theorie-Praxis-Debatte in der modernen deutschen und amerikanischen Soziologie* (Reinbek bei Hamburg: Rowohlt).

Beck, Ulrich, 1993: *Die Erfindung des Politischen* (Frankfurt: Suhrkamp).

Bell, Danie, 1960, 1966: *The End of Ideology* (New York: Free Press).

Bell, Daniel, 1993: *Communitarianism and Its Critics* (Oxford: Clarendon).

Beyme, Klaus von, 1988: "Die Rolle der Theoriegeschichte in der modernen Politik-wissenschaft: das Beispiel Amerikas", in: Beyme, Klaus von (Ed.): *Der Vergleich in der Politik-wissen-schaft* (Munich: Piper): 88–108.

Beyme, Klaus von, ^{3}1996, ^{4}2007: *Theorie der Politik im 20. Jahrhundert. Von der Moderne zur Postmoderne* (Frankfurt: Suhrkamp).

Beyme, Klaus von; Offe, Claus (Eds.), 1996: "Politische Theorien in der Ära der Transformation, in: *PVS Sonderheft*, Vol. 26. (Opladen: Westdeutscher).

Bobbio, Norberto, 1989: *Democracy and Dictatorship* (Cambridge: Polity).

Cohen, Jean L.; Arato, Andrew, ^{3}1995: *Civil Society and Political Theory* (Cambridge, Mass.: MIT Press).

Cohen, Joshua; Walzer, Michael, 1986: "Spheres of Justice" (Book Review), in: *The Journal of Philosophy*, 83: 457–468.

Cohen, Joshua; Rogers, Joel, 1992: "Secondary Associations and Democratic Governance", in: *Politics and Society*, 4 (special issue): 393–472.

Cohen, Joshua; Rogers, Joel, 1994: "Solidarity, Democracy, Association", in: Streeck, Wolfgang (Ed.): *Staat und Verbände*. PVS special issue 25 (Opladen: Westdeutscher): 136–159.

DeLeon, P., 1993: "Demokratie und Policy-Analyse", in: Héritier, Adrienne (Ed.): *Policy-Analyse*. PVS special issue 24 (Opladen: Westdeutscher): 471–485.

Downs, Anthony, 1957: *An Economic Theory of Democracy* (New York: Harper & Row).

Etzioni, Amitai, 1993: *The Spirit of Community* (New York: Crown Publication).

Gerecke, Uwe, 1995: "John Rawls' Beitrag zu einer Theorie der modernen Gesellschaft", in: Pies, Ingo; Leschke, Ingo (Eds.): *John Rawls' politischer Liberalismus* (Tübingen: Mohr): 21–47.

Goodin, Robert E., 1992: *Motivating Political Morality* (Oxford: Blackwell).
Habermas, Jürgen, 1981: *Theorie des kommunikativen Handelns*, Vol. 2 (Frankfurt: Suhrkamp).
Habermas, Jürgen, ^{2}1988: "Entgegnung, in: Honnet", Axel; Joas, Hans (Eds.): *Beiträge zu Jürgen Habermas' Theorie des kommunikativen Handelns* (Frankfurt: Suhrkamp): 327–405.
Held, David, 1989: *Political Theory and the Modern State* (Cambridge: Polity).
Honneth, Axel, 1992: *Kampf um Anerkennung. Zur moralischen Grammatik sozialer Konflikte* (Frankfurt: Suhrkamp).
Huntington, Samuel, 1981: *American Politics: The Promise of Disharmony* (Cambridge: Harvard University Press).
Kersting, Wolfgang, 1996: *Die politische Philosophie des Gesellschaftsvertrags* (Darmstadt: Primus).
Kymlicka, Will, 1990: *Contemporary Political Philosophy* (Oxford: Clarendon).
Lindner, Clausjohann, 1990: *Kritik der Theorie der partizipatorischen Demokratie* (Opladen: Westdeutscher Verlag).
Luhmann, Niklas, 1993: *Das Recht der Gesellschaft* (Frankfurt: Suhrkamp).
MacIntyre, Alasdair, 1981: *After Virtue* (Notre Dame: Notre Dame University Press).
Mannheim, Karl, 1952: *Ideologie und Utopie* (Frankfurt: Schulte-Bulmke).
Moore, Barrington, 1978: *Injustice. The Social Bases of Obedience and Revolt* (London: Macmillan).
Naschold, Frieder, 1969: *Organisation und Demokratie* (Stuttgart: Kohlhammer).
O'Neill, Onora, 1996: *Towards Justice and Virtue* (Cambridge: Cambridge University Press).
Peters, Bernhard, 1993: *Die Integration moderner Gesellschaften* (Frankfurt: Suhrkamp).
Philips, D. C., 1976: *Holistic Thought in Social Science* (Stanford: Stanford University Press).
Ricci, David M., 1984: *The Tragedy of Political Science. Politics, Scholarship and Democracy* (New Haven: Yale University Press).
Rödel, Ulrich et al., 1989: *Die demokratische Frage* (Frankfurt: Suhrkamp).
Sabine, George, 1989: "What Is Political Theory?", in: Gould, J. A.; Thursby, V. V. (Eds.): *Contemporary Political Thought* (New York: Holt Rinehart & Winston): 7–20.
Sartori, Giovanni, 1994: *Comparative Constitutional Engineering* (Houndsmill: Macmillan).
Schmalz-Bruns, Rainer, 1995: *Reflexive Demokratie* (Baden-Baden: Nomos).
Voegelin, Eric, ^{2}1965: *Die neue Wissenschaft der Politik* (Munich: Pustet).
Walzer, Michael, 1983: *Spheres of Justice* (New York: Basic Books).

Chapter 6
Theodor W. Adorno: Political Theory as Theory of Aesthetics

6.1 Introduction[1]

Adorno's 100th anniversary is more difficult to celebrate than the anniversaries of other thinkers. In the early days there was a collective enterprise of thinking with Horkheimer, who in 1930 became the director of the "Institute for Social Research" in Frankfurt. Sometimes the 'twin thinkers' were compared to Marx and Engels to whom both owed a lot. In the tradition of Hegel both wanted to overcome the division of philosophy and empirical research in a critical philosophy of history. The Frankfurt school in its first phase still defined history as a process of the progress of productive forces, but from the outset they were convinced that these productive forces were neither the incarnation of rationality, nor that the proletariat was necessarily the torch-bearer of social progress. Adorno, however, was more than Horkheimer convinced that 'critical theory' could no longer be a form of reflection for the organised working class.

Adorno was more than the other important members of the Institute—such as Marcuse, Fromm, Kirchheimer and others—struck by the experience that the rise of fascism undermined the belief in the rationality of a development of technology and civilisation. Insofar Adorno was—as many right wing thinkers from Heidegger to Arnold Gehlen in Germany—in the tradition of the intellectual dichotomy of 'civilisation' and 'culture'. Anglo-Saxons and French writers on the basis of their languages had no chance to denigrate 'civilisation' and to hail 'culture' in quasi metaphysical terms as the incarnation of totality and the unity of rational and emotional values. But his criticism of technocratic society and the search for true culture distanced itself from the conservative thinkers of his time, such as Jaspers, Ortega y Gasset and even Huxley who saw culture as a value per se. True enlightenment does not conserve the cultural values of the past but fights for the lost hopes of former generations (DdA: 15).

[1] This text was presented in Mexico City in 2003 at the National Autonomous University of Mexico (UNAM) and has not been published previously.

K. von Beyme, *Klaus von Beyme*, SpringerBriefs on Pioneers in Science and Practice 14,
DOI: 10.1007/978-3-319-01535-4_6, © The Author(s) 2014

Though there is no unified 'Frankfurt school' and even Adorno and Horkheimer differentiated in their work, the younger generation of the Frankfurt school, such as Wellmer (1985: 137) thought that it was almost impossible to evaluate Adorno's political philosophy. Adorno avoided systematic thinking. Every dogmatic fixation was part of the over-all critique of the reality of his time. Adorno was always afraid of the ordinary scientific language. This is why he preferred the use of foreign language notions (O'Neill 1999: 81) which—combined with a very individualistic use of his German language—makes his work not easy interpretable. Adorno hated the academic philosophy in which he had trouble to be successful and which he blamed to act according the opposite device of Descartes' famous slogan: "Sum, ergo cogito" (In a free translation: I am sitting on an academic chair and thus I have to think). Adorno distrusted general notions, but nevertheless was not consequent enough to turn to poetry of music and thus did not avoid 'reification' of many notions which became political slogans in 1968. As an empirical scholar I think we cannot avoid generalising his ideas and subsuming them under certain general notions—via comparisons.

Adorno's way of thinking was certainly influenced by his early hopes of becoming a musical artist which he abandoned only about 1925 and tried to get his second doctorate (Habilitation) in philosophy (1927) (Wiggershaus 1986: 98). More than other members of the Institute—who like Pollock still worked on a theory of 'state monopoly capitalism'—Adorno entered into polemics with the narrow rationality concepts of Marxism. Under the influence of Walter Benjamin he added two new approaches to Marxist philosophy:

- The synthesis of Freud and Marx, an interest shared with Marcuse and Fromm,
- the emphasis on aesthetic interpretation of civilisation and culture.

The unique development of Adorno combined leftist Hegelian thought with the cultural pessimism of thinkers such as Schopenhauer and Nietzsche who were mainly adopted by right-wing philosophy. With Horkheimer, Adorno worked in the early 1940s on the book on *Dialectics of Enlightenment* which developed a negative philosophy of history. Under the impression of emigration and persecution by the Nazis they discovered a 'logic of decay and disintegration'. In the process of civilisation, 'Verdinglichung' (reification) seemed to be the fate of a social development which was built on the mere domination of nature. This process—built on a system's theory—has neglected the communicative values of the sphere of life, an idea which later was developed by Adornos assistant Jürgen Habermas. Whereas Habermas, however, later came increasingly to a new synthesis of a system's and a sphere of life perspective, Adorno saw no chance to develop political possibility for his social research and for concrete political practice. At best the result was a kind of 'attentism'—in a period of messianic hope for an interruption of the process of civilisation. Under the influence of Benjamin the critique of existing societies included social democratic devices for incremental changes and developments toward a more human socialist society.

6.2 Studies in Fascism

Max Horkheimer even during the war continued to hope that 'active resistance' against fascism was possible. Contrary to Bert Brecht (*Der unaufhaltsame Aufstieg des Arturo Ui*) and other German left-wing writers Adorno and Horkheimer did not consider the Nazis as a mere clique of gangsters, usurping power via coup d'état. According to the social psychological approaches they recognized the correspondence of the Nazis with the (however mislead) intentions of great parts of the German society. Adorno (MM: 120) came up with the odd idea that even Hitler was 'too liberal' to recognise all the failures of liberalism which he hated. Fascist leaders were 'Charaktermasken', victims as well as criminal actors. Hitler himself was hardly more than the 'drummer of his society'

When Adorno and his collaborators worked on the project of 'Studies in Prejudice', sponsored by American Jewish Committee, they overcame a former prejudice of their own: that capitalist countries such as Nazi Germany and democratic America were unlikely to wage a war against each other. Fascism was the natural product of a crisis of international capital (Doc. in: Dubiel 1978: 62ff). *The Authoritarian Personality*, part one of the studies led by Adorno, Else Frenkel-Brunswik, Daniel Levinson and R. Nevitt Sanford, was developed in a time when the USA fought against fascism and was allied to the Soviet Union (Adorno et al. 1968, vol. 2).

The empirical work with surveying more than 2,000 students was done, however, in a time of the cold war, when the hopes for a new human 'pax Americana' in the whole world seemed to fail. The fascist character was no longer something which existed only in Europe. Fascist characters were—contrary to the current American folklore—nothing particularly inherent in the Germans. That Italians and Germans first became pray of fascist regimes was rather explained in Marxist terms, with the traditional character of capitalism in these countries.

Fascist characters, however, could also develop in higher developed capitalist societies. They were characterised by rigid conventional middle-class values, conformist behaviour, combined with pessimistic anthropology, suspicion against sexual freedom, submission under hierarchies and idealised authorities, combined with hatred against minorities. In terms of psychoanalysis the fascist character in the Freudian trinity showed an 'underdeveloped ego', an externalised 'super-ego' and an ego-hostile 'it' (Es/it). Adorno wrote in a chapter on the AP project that his aim was a cultural anthropology of anti-semitism and "to find out how objective economic laws operate, not so much through the individual's economic 'motivations' than through his unconscious make-up…" (Remarks on the Authoritarian Personality: 15; cit. Wiggershaus 1986: 478).

The studies in prejudice about authoritarian personalities was a progress compared to the anthropological literature of that time which explained authoritarian attitudes by some kind of national character. Odd generalisations were used such as a causal links between methods of child rearing in Germany, Japan and Russia and the respective political systems (Gorer, Mead and others). Most influential was the fascism scale, but it was never used in an independent survey. Empiricists criticized that the theoretical explanations had little to do with the empirical studies. Authoritarian character was explained by a sadomasochist solution of the Oedipus complex, without indicating clear causalities. Positivist thinkers like Hempel (1965: 297ff) compared this kind of procedure with another metatheoretical position working with the assumption of 'totalities", e.g., functionalism.

By Adorno the jew is considered as the substitute of the hated father. Sadism is used against alien groups. Different groups of potentially authoritarian characters were classified such as rebels, rowdies, grumblers, manipulated persons, characterised by thinking in stereotypes. This latter group persecutes alien groups without hatred. Persecution for them is a mere problem of organisation. Concentration camp personnel seems to fit this type of fascist personality. The normal unauthoritarian character was the incarnation of what Adorno's critical theory was aiming at: the protesting person without prejudices who is not to be confounded with the laisser-faire-type or the manipulative type without prejudices. Socialisation and working place conditions were mostly not dealt with, though Adorno in that period tried to pass as a sociologist and not only as a philosopher. Language problems were pondered over as far as the definition of concepts is concerned. The language barriers of lower class persons, however, were not taken seriously enough. Empiricists criticised therefore the premature concept of working class authoritarianism (Jaerisch 1975: 156). Empirical social psychologists in America, such as Edward Shils (1954: 23ff) criticised that Adorno and his team were unable to explain why certain authoritarian groups did not respond to Adorno's questions in the provided way. The differences between right and left wing extremists were blurred. The psychologist Eysenck (1968: 206ff) was more open to see differences between fascists and communists and stated a certain 'tendermindedness' among Communists concerning many of the questions. Only in the time of the student's rebellion the differences were blurred against. Even Habermas had to resent later that he once—under pressure from the students—spoke about 'leftist fascism'. Critical theory avoided clearly defined causalities by epitheta such as 'tendentially" in order to hint at processes which were likely to develop in the future—though the empirical analysis of the present time did not offer a clear picture.

This made it possible to compare fascists, communists and pseudo-democrats in America and to see 'a tendency' of rising authoritarianism everywhere in the world. Criticism against Roosevelt's administration seemed to indicate latent pseudo-conservatism, likely to develop into authoritarianism. McCarthyism was welcome to justify this whole-sale condemnation of American developments after the war. Adorno's criticism of the commodity society and the culture industry seemed to be directed against America. It helped to strengthen anti-american feelings of the rebelling students in the time of the Vietnam War. When during the Iraque crisis Anti-Americanism was suspected again in Germany it was good that diligent researchers had found a pro-American text by Adorno which he never had published. In this presentation America was a positive society with its friendliness—though trained in 'charm schools', its extroverted human relations and its abundance of goods like in Paradise (Adorno 2003).

The feeling of alienation in the USA was one of the reasons why Adorno and Horkheimer were able to return soon after the war to Germany. Californian friends like Thomas Mann—for whom Adorno had served as an adviser for the musical theory his book on *Dr. Faustus* and Fritz Lang—did not approve this decision. It was even noteworthy that Adorno entered into an almost friendly correspondence with a conservative thinker such as Arnold Gehlen who had worked with the Nazis in Germany (Thies 1997: 50) and after the war criticised modern culture of the Avant-garde from the right, still using older elitist terms such as 'Vermassung' (Mass society). Whereas the student's who followed the Critical School around 1968 discovered former fascists everywhere, there was a great resistance in Adorno's statements to denounce the political past of other theoreticians, even including Heidegger whom he fought as the arch-priest of an 'essentialist jargon' (Jargon der Eigentlichkeit) (JdE). Only discretely Adorno worked against hopes to get at chair for Gehlen at Heidelberg University.

Empirical research later did not stick to the concepts developed by the Frankfurt team in America. It showed frequently that 'deviant political behaviour' may be rational though it seems to be irrational. Empirical research in violence has overcome many of the prejudices in the "Studies of prejudice". The studies on the authoritarian personality made it clear even for the authors such as Adorno that their main interest was concept building and critical theory of society. The fruit of this inside was Adorno's and Horkheimer's seminal study on *Dialectic of Enlightenment*, written in the early 1940s.

6.3 Political Theory as Theory of Aesthetics

Political theory as theory of aesthetics had a deep tradition in German intellectual history. Engels—after the death of Marx—was approaching already the idea that the cultural superstructure was not strictly determined by economic processes. Already Engels recognised a gap between the declared intentions of an artists and the objective outcome which was more than the result of class origins. 'Cultural Marxism'—in the tradition of Lukács, Korsch or Gramsci—developed a tradition in different countries. In France it was represented by Sartre and Lucien Goldman, in America by Edmund Wilson and Sidney Finkelstein.

In the German speaking literature the Hungarian philosopher Georg Lukács—mostly writing in German—tried to bridge the gap between Engels position and Lenin's demand for class-bound arts and literature. Lukács developed the division between realism and naturalism. Naturalist realism with a progressive tendency as in the work of Zola was in this perspective less interesting than the analysis of societies by conservative writers from Balzac to Walter Scott. The Frankfurt school, Ernst Bloch and Walter Benjamin, remained faithful to their emphasis on culture as the basic concept of social development, whereas Lukács after the orthodox Marxist criticism of his book 'History and class consciousness' which was considered as a continuation of idealism under the disguise of Hegelian political philosophy. Lukács—who under Béla Kun 1919 and under Imre Nagy 1956 twice served as minister of culture and education—remained more interested in practical politics and transitorily made concessions to Marxist orthodoxy. The differents developments of Adorno and Lukács were undoubtedly influenced by their favoured country of emigration. Adorno went to the USA—Lukács lived for a long time in the USSR.

Critical theory was developed in opposition with Kant's definition of the beautiful as 'pleasure without interests' (interesseloses Wohlgefallen). True art is lead by the legitimate interest of man in a future happiness. Aesthetic experience contains a progressive element of 'common humanity'. This meta-individual subject is not—as in Kant—an abstract transcendental entity but a concrete historical subject. This was one of the reasons why critical theory was critical of the dominant current in arts and literature in the Weimar period, e.g., expressionism with its excessive subjectivism. Aesthetic theory as critique of the existing society and the manipulated language of the dominant classes in it was for Adorno necessarily political. But the negative dialectical approach never allows under the existing condition of distortions of the subject and its rationality and the alienation of mankind a complete synthesis of the existing and the values of a future better society. Art became autonomous. All attempts to restore a social and political function of art he considered as a failure (ÄT: 9). Cultural industry with its unifying character which annihilates differences in architecture, movies or literature was the archenemy of critical theory from the outset (DdA: 141ff), because it extinguishes individuality. Culture thus seemed to be restricted to a duplication of a miserable social reality. Amusement was the correlate of being bored. Even Chaplin's Anti-Hitler-movie seemed to be but a duplication of boring and flat Nazi every-day life (DdA: 171).

Sociology of art for Adorno had very limited functions. Mozart's social conditions did not really explain his music. But the contrary declaration of the 'l'art pour l'art"-movement in his theory contained also deceptions because it was not able to avoid that art became a 'commodity' and even 'Kitsch' (ÄT: 352). Though he recognized in his work on the 'Jargon of essentialism' (Jargon der Eigentlichkeit' that the German symbolists such as George and Hofmansthal were not writing jargon, their philosophical and literary interpreters down to Heidegger were open to the fallacy of jargon (JdE: 417).

Adorno's dialectic distanced itself from Hegel's Aesthetics. He agreed with Hegel only on one point: aesthetics cannot be separated from the rest of social reality but permeates it and is not just another branch of thinking (cf. Specht 1981:114). Adorno liked encyclopaedic work without specialisation and hoped for the end of the limitations to various arts. The end of the 'genres› and the 'styles' was proclaimed (Eichel 1993: 20ff). 'Interlacing'—which he called 'Verfransung'—was his ideal. This was one reason why he had a lot of understanding for avantgardist 'montage' and 'assemblage'.

Hegel's 'cunning of reason' (List der Vernunft) in Adorno's philosophy got a negative equivalent: 'the cunning of unreason' (ÄT: 331). Art was political in opposition against the existing system. No static style or aesthetic doctrine was accepted. Hegel and Marx were accused of their open classicist attitude (ÄT: 309). The strange synthesis of ideas in the tradition of Hegel and Marx on the one hand and of Schopenhauer and Nietzsche on the other hand, led to a preponderance of the latter tradition. Nietzsches emphasis on the 'dionysic' was transformed into political progressivism. In music the original sin of deterioration was the overcoming of polyphony and the doctrine of the 'thorough bass' (Generalbass) with its hierarchisation of musics (ÄT: 311). The ontological classic approach was for Adorno (ÄT: 301) characterised by an 'intolerance of ambiguity'. This view made Adorno so interesting for postmodern thinking and explains why a failed political philosopher is honoured in all parts of the world at his 100th anniversary.

Adorno exemplified his ideas particularly in music and literature. He did not accept the difference of 'serious music' and 'light music'. Adorno was not a traditionalist and 'Bildungsbürger' (cultural bourgeois) defending classical art. In his Aesthetic theory (ÄT: 19ff) his differentiation hinted at arts aiming at the 'market' and arts transcending the dominant manipulated 'false consciousness'. Marx's theory of alienation in his 'economic period' had developed into a theory of 'commodity fetishism' (Warenfetischismus) and Adorno showed this tendency in modern art, music and literature. Contrary to the Leninist and Stalinist development of Marxist aethetics into a banal 'socialist realism' Adorno's idea of a 'vanguard' aimed at an art which transcended the tastes and intentions of the labouring classes. This was the reason why Adorno was one of the first theoreticians of music who hailed Arnold Schönberg's development of dissonances not finding a synthesis in a new harmony (Pr: 180ff). Negative dialects in arts meant a refusal to find compromises in arts which reflected the unsolved conflicts in society. "Art is the social antithesis to society" (ÄT: 29).

Adorno's theories by no means were free from errors. During his stay in Britain he wrote on 'Jazz' under a pseudonym of 'Hektor Rottweiler' (ZfS 1936, V: 2 reprinted in 'Moments musicaux': 88). Jazz in this period was the perfect expression of commodity fetishism. By some critics he was even suspected that he wanted to please the fascists with his criticism of Jazz which was considered by them as the expression of an inferior—mostly black—race. Black contributions to Jazz were considered by Adorno not as a rebellion against the racist society but as compliance with it. Jazz was part of the pseudo-culture of entertainment, not aiming at active-subjective practice but at masochist passivity. Jazz was a mechanical and ritualistic answer by 'musically illiterates' to an alienated world. 'Kunst wird entkunstet'—a game on worlds hard to translate: art looses its art character by offering a way of adaptation to the repressive capitalist society (Pr.: 159). Only in the new gender relations there was a positive element in the culture of Jazz. Adorno never abandoned this analysis which showed a limited capacity to learn from the society which offered him refuge in the USA. Even in a contribution on 'a fashion beyond the time' (1953 'Zeitlose Mode—zum Jazz' (Pr.: 144ff) and shortly before his death he endorsed former views and called them 'still too optimistic'. During all his life Adorno emphasised that 'fetishisation' was not only a psychological category but also an economic notion, denouncing an art which was not guided by its 'value of use' (Gebrauchswert) but by its commercial 'change value' (Tauschwert) in a market society. These contributions fitted into the critical approach against the radio which support a depersonalised, collective and alienated form of listening, without developing the negative function of critique which was essential to his political theory as a theory of aesthetics.

Adorno's work is not only full of contradictions and unjust evaluations in his domain of music. The pioneer of modern music had no understanding for modern architecture which he lumped altogether into the rubric of anti-human megalomania (DdA: 141ff)—in spite of his love for the 'Verfransung' of various arts. This negative attitude towards modern architecture made him, however, popular in the post-modern period of feminist and postmodern criticism of the classical avant-garde in architecture (O'Neill 1999: 174ff).

As in his studies on radio diffusion and fascism this unempirical approach brought Adorno to an increasing disagreement with the Austrian emigré and pioneer of empirical social research Paul Lazarsfeld. Adorno's transfer from New York to Los Angeles in 1941 brought the always difficult cooperation with Lazarsfeld to an end. The cooperation with Lazarsfeld was caused by the "Princeton Radio Research Project". Lazarsfeld originally was eager to co-opt the author of renowned writings on music. But from the outset the divergences became evident: Adorno emphasised theory, claiming that this did not mean neglecting empirical work. Lazarsfeld (Adorno 3.2.1938) accepted the importance of theory but concluded: "On the other hand, we shall have to understand that you have to end up finally with research among listeners, although in many cases we might have to stop with the formulation of the theoretical problem and discussions of techniques to answer them, simply for reasons of time" (cit. Wiggershaus 1986: 268). Later quarrels of the critical theory against 'positivism' had an early reason in disagreements between European emigrés. When Adorno tried to work empirical Lazarsfeld's reaction was highly negative: he called him 'uninformed about empirical research' and criticised that Adorno did not 'exhaust the logical alternatives' of his own statements. Finally he personalised his reproaches in a very blunt way: "You attack other people as fetishists, neurotic and sloppy but you show yourself the same traits very clearly' (ibidem: 272). The hopeful synthesis of theory and empirical research in the studies on Authority and Family petered out and the division of labour between theory and empirical research became unavoidable.

6.4 Philosophical Work on 'Negative Dialects'

After disappointments over empirical work 'critical theory' in Adorno's and Horkheimer's work developed into a critique of 'Enlightenment' which was considered as a continuation of mythology under the disguise of rational theory. The fact in Adorno's theories that fascism was nothing inherent in the German national character per se made it possible for him to return to Frankfurt. His main explanation was that he needed the language. Indeed his style of writing was close to literature and philosophical essayism. *Minima moralia* (1951, 2003) were certainly the classical example of Adorno's way of reasoning. It was a kind of 'founding document' of the post-war Frankfurt school and after the experiences of totalitarianism the theory of an abused modernity and mislead avant-garde which emphasised a 'universal syndrom of illusion' (universeller Verblendungszusammenhang) which transcended all the former theories of ideological thinking from Karl Marx to Karl Mannheim. His late work on *Negative Dialects* was the elaboration of this initial common venture with his friend Horkheimer.

The main enemy was positivist thinking and the remainders of 'affirmative statements' even in critical thought (ND 2003: 9). Only from the standpoint of Messianism a theory of authentic rationality and 'Vernunft' and a denunciation of 'false consciousness' seemed to be possible. Negativism and Messianism were somehow combined.

But the old Jewish 'anti-image-verdict', the 'Bilderverbot'—proclaimed already by Marx against the utopian Communists of his time like Cabet, Fourier or Weitling—was still valid in Adorno's work. Utopian thinking for Adorno (MM: 299) was mainly a copy of those societies against which it pretended to be directed.

Dialects in ancient Greece developed among the sophists as a method to find the truth. Even in Hegel there was a 'synthesis' and in Marx the was a 'negation of the negation' as a third step of reasoning. Critical theory does not offer any "positive ideas of a right society" (MM: 299). Dialectics had to be permanently defended against abuse so that it could fight ideas who claimed to be right and to offer asylum to the ideas of the suppressed.

Only philosophic thought is a momentum for revolutionary changes—not organised violence by self-appointed spokesmen of the proletariat. There is, however, a kind of vanguard, the intellectuals. Against Bender's hypothesis of the "trahison de clercs" Adorno (MM: 302) hoped that the masses no longer distrusted the intellectuals because of a treason of revolution, but rather because they were honest in promoting intellectual revolution. Adorno was less disappointed with Marxism than his friend Horkheimer because he never believed essential theories of revolutionary Marxism. Whereas Marx in his famous 11th thesis on Feuerbach emphasised that philosophers so far have only interpreted history in various ways—but the task of the future was to change it—Adorno came back to mere interpretation.

Adorno's basic experience—in tune with one tradition of German thinking from Nietzsche to Heidegger—was the experience of the non-sense of a life into which men were thrown against the will. 'Geworfenheit' was a basic concept in Heidegger's philosophy. Adorno's main foe was 'das Ganze', the whole, a totality, thought in terms of a non-contradictory scientific entity. He suffered from a neurotic sense of senselessness. But at the same time—in the light of Freud—he derived from this a 'secondary gain from bad health' (sekundärer Krankheitsgewinn). His drug was mainly work, and his obsessive concern for thinking—sometimes in harsh judgements over other thinkers—did not make him popular. Normative thinkers, such as Hannah Arendt, hated him. The musician Schönberg who owed a lot to Adorno's theories did not like him. Even with his friend Horkheimer he was several times at odds. The last time was Horkheimer's refusal to accept Adorno's assistant Jürgen Habermas (Doc. in: Claußen 2003: 407–419). Today it is hardly understandable in the light of Habermas' development that he was considered as 'too revolutionary' by Horkheimer. Even friends, such as Siegfried Kracauer resented his 'vague profoundness' ('ausgeleierter Tiefsinn') which made it possible that dialectics led to so many contradictory statements in Adorno's work.

6.5 A Vision of Good Society?

Most controversial was the existence of a normative political theory in Adorno's work. Most positivists, including this writer, subsumed critical theory among 'normative political philosophy'. But contrary to Strauss, Voegelin, Arendt an other conservative thinkers from German speaking countries, there was only a negative notion of what should be in Adorno's work. A negative theology has frequently been stated by the literature (Deuser 1980; Koch/Kodalle 1973; Wischke 1994: 153). Even post-war democracies were not 'open' enough. 'After Auschwitz' nothing made sense, even not art and poetry. There was no real life for Adorno in a 'misconceived totality'. Postmodernists have compared Adorno in this respect even with Woody Allen (Witkin 2003: 157). But a kind of theological element was present in the hope for salvation via thinking and via the arts. A free society was never—as in Marxism—a society of rational development of the productive forces—but always a society which realised 'humanism' without exploitation, competition struggles and de-individualisation of isolated subjects.

A school of 'Critical Theory' in Germany was recognised only when Adorno entered the controversies about 'positivism'. This debate took place more in sociology than in philosophy. In 1961 Adorno (1970: 108ff) was the main antagonist against Karl R. Popper and his version of what Adorno lumped together as 'positivism'—which for itself stole the epitheton 'critical' in dubbing the group as 'critical rationalism'. Adorno and Popper were at one only in the postulate of a primacy of theory—which separated critical rationalism from ordinary positivism. Positivism for Adorno was an alienation of thinking from social reality, a logic of invented 'tokens for a game' with data. Adorno universalised suspicion against ideologies—technocracy—hailed by conservative thinkers such as Gehlen as a way to overcome ideologies—and science, both for Adorno were predominantly new forms of ideology. This universal suspicion against ideology later by Habermas (1988: 156) was recognised as 'undialectical' and almost theological, because a kind of devil seemed to haunt social reality which had to be exorcised by Critical Theory.

Critical Theory which wanted to stay critical needs a normative basis outside the description of the pathologies in society. There was a dilemma, however. If the 'anti-authoritarian principle" has to transform itself into an institution against reason (DdA: 112), than no norms can be founded in a logical way. Norms seem to be contingent. This conclusion—drawn by Nietzsche and the Marquis de Sade—was, however, not accepted by Adorno. Critical theory thus remained within the framework of idealism and did not completely live up to its own expectation of being "anti-essentialist" (Bonacker 2000: 276). Max Weber's way of reasoning, to create institutions because modern polytheism does not allow one logical hierarchy of norms, seems to be more viable for a modern theory of society. But institutions for Adorno are not acceptable, not even in Weber's soft version that institutions have to remain open in order to recreate daily a new balance between normative contingency and the need for an institutional stability in society. Adorno showed—contrary to Horkheimer—not even an interest in his own institution, the Institute for Social Research when he had do lead it. According to his former assistant Habermas (2003: 45) Adorno was only 'the passive centre' of cross-cutting pressures from his wife, Gretl, the emeritus Horkheimer and his co-director von Friedeburg.

Adorno was particularly unsatisfied with the way conservative thinkers, such as Gehlen, used institutions as a substitute for the lost instincts of mankind. Whereas Popper and his followers built on Weber, Adorno fought the 'mandarins of modernisation' in German history of ideas.

Adorno's aloofness from concrete political reasoning made his work hardly usable for political science. He was not able to explain the failure of the Weimar Republic nor the problems of post-war democracy. When the student's movement took literally many assumptions of Adorno's and demanded a return to the 'revolutionary origins' of the Frankfurt School, Adorno was deeply alienated and withdrew into his philosophical ivory tower. Some critics thought that the conflict with the students who had loved Adorno and now turned against him contributed to his early death (Claußen 2003: 381). Nevertheless Adorno remained politically 'responsible'. He was not an escapist thinker. His engagement for the 'unidentical' created an extreme sensibility for injustice in society.

Critical theory has overcome this dilemma only after the 'Weberisation' in the work of Jürgen Habermas. The negative forces of life were now called 'system'. But the face-to-face world of life (Lebenswelt) could be defended against the intrusion from the system with its negative forces of bureaucratisation, judicialisation and commercialisation. Not all organisation was negative in this revised version of critical theory. The new social movements became currents of hope for enlightening 'the system' via critical thinking and action. Habermas (in: Honneth/Wellmer 1986: 12) admitted that the hope of creating an interdisciplinary theory of society in the Max-Planck Institute at Starnberg—which he directed with the philosopher and physicist Carl Friedrich von Weizsäcker—failed. But the theory of communicative action was indeed the most fruitful result of the aspirations of the Frankfurt School. Habermas overcame Adorno's bias of dealing predominantly with cultural industry. The means of mass communications had developed and were worked into a modernised version of Critical Theory.

References

Sources

Adorno. Th. W. et al., 1950, 1968: *Studies of Prejudice*, Vol. 2 (New York/Amsterdam: De Munter).

Adorno, Th. W., 1955: *Prismen: Kulturkritik und Gesellschaft* (Frankfurt: Suhrkamp). (quoted as P).

Adorno, Th. W., 1970: *Aufsätze zur Gesellschaftstheorie und Methodologie* (Frankfurt, Suhrkamp).

Adorno, Th. W.: 1986ff: in: Tiedemann R. (Ed.) *Gesammelte Schriften*, Vol. 20 (Frankfurt, Suhrkamp) (GS).

Adorno, Th. W.; Benjamin, W., 1994: in: Lonitz H. (Hrsg.): *Briefwechsel 1928-1940* (Frankfurt: Suhrkamp).

Adorno, Th. W., 2003: "Mister Amerika", in: *Süddeutsche Zeitung Magazin*, 29 August: 18f (an unknown paper as a praise of the American way of life).

Adorno, Th. W., 2003: in: Tiedemann R. (Ed.): *Hauptwerke*, Vol. 5 (Frankfurt: Suhrkamp).

Adorno, Th. W.; Horkheimer, M.: *Dialektik der Aufklärung*. GS, Vol. 3 (quoted as: DdA).

Adorno, Th. W.: *Negative Dialektik. Jargon der Eigentlichkeit*. GS. Vol. 6 (quoted as: ND - JdE).

Adorno, Th. W.: *Ästhetische Theorie*. GS Vol. 7 (quoted as: ÄT).

Adorno. Th. W.: *Philosophie neuen der Musik*. GS. Vol. 12 (quoted as: Ph d.n.M).

Literature

Always, J., 1995: *Critical Theory and Political Possibilities. Conceptions of Emancipatory Politics in the Works of Horkheimer, Adorno, Marcuse, and Habermas* (Westport: Greenwood).

Apel, H., 1980: *Die Gesellschaftstheorie der Frankfurter Schule* (Frankfurt: Diesterweg).

Becker, M., 1997: *Natur, Herrschaft, Recht. Das Recht der ersten Natur in der zweiten: Zum Begriff eines negativen Naturrechts bei Theodor Wiesengrund Adorno* (Berlin: Duncker & Humblot).

Bonacker, T., 2000: *Die normative Kraft der Kontingenz. Nichtessentialistische Gesellschaftskritik nach Weber und Adorno* (Frankfurt: Campus).

Breuer. St., 1992: *Die Gesellschaft des Verschwindens. Von der Selbstzerstörung der technischen Zivilisation* (Hamburg: Junius).

Claußen, D., 2003: *Theodor W. Adorno. Ein letztes Genie* (Frankfurt: Fischer).

Deuser, H.,1980: *Theologische Theologie: Studien zur Adornos Metaphysik und zum Spätwerk Kierkegaards* (München).

Dubiel, H., 1978: *Wissenschaftsorganisation und politische Erfahrung. Studien zur frühen Kritischen Theorie* (Frankfurt: Suhrkamp).

Eichel, Ch., 1993: *Vom Ermatten der Avantgarde zur Vernetzung der Künste. Perspektiven einer interdisziplinären Ästhetik im Spätwerk Theodor W. Adornos* (Frankfurt: Suhrkamp).

Eysenck, J. ^{5}1968: *The Psychology of Politics* (London).

Habermas, J., 1985, ^{4}1988: *Der philosophische Diskurs der Moderne* (Frankfurt: Suhrkamp): 130–157.

Habermas, J., 2003: "Die Zeit hatte einen doppelten Boden. Der Philosoph Theodor W. Adorno in den fünfziger Jahren", in: *Die Zeit*. 4 September: 45f.

Hempel, C. G., 1965: *Aspects of Scientific Explanation* (New York: Free Press).

Honneth, A.; Wellmer, A. (Eds.), 1986: *Die Frankfurter Schule und die Folgen* (Berlin: de Gruyter).
Jaerisch, U., 1975: *Sind Arbeiter autoritär? Zur Methodenkritik politischer Psychologie* (Frankfurt: EVA).
Jay, M., 1973: *The Dialectical Imagination. A History of the Frankfurt School and the Institute of Social Research 1923–1950* (Boston: Little Brown).
Kilian M., 1998: *Modern and Postmodern Strategies. Gaming and the Question of Morality* (New York: Lang).
Koch, T.; Kodalle, K.-M., 1973: *Negative Dialektik und die Idee der Versöhnung* (Stuttgart).
Lunn, E., 1982: *Marxism and Modernism. An Historical Study of Lukács, Brecht, Benjamin, and Adorno* (Berkeley: University of California Press).
Müller-Doohm, St., 2003: *Adorno'. Eine Biographie* (Frankfurt: Fischer).
O'Neil, M. (Ed.), 1999: *Adorno. Culture and Feminism* (London: Sage).
Paetzel, U., 2001: *Kunst und Kulturindustrie bei Adorno und Habermas* (Wiesbaden: Deutscher Universitätsverlag).
Shils, E., 1954: "Authoritarianism: Right and Left", in: Christie, R.; Jahoda, M. (Eds.): *Studies in the Scope and Methods of 'The Authoritarian Personality'* (Glencoe, Ill.: Free Press): 23ff.
Specht, S., 1981: *Erinnerung als Veränderung. Über den Zusammenhang von Kunst und Politik bei Theodor W. Adorno* (Mittenwald: Mäander).
Thies, Ch., 1997: *Die Krise des Individuums. Zur Kritik der Moderne bei Adorno und Gehlen* (Reinbek: Rowohlt).
Wellmer, A., 1985: *Zur Dialektik von Moderne und Postmoderne. Vernunftkritik nach Adorno* (Frankfurt: Suhrkamp).
Wiggershaus, R., 1986: *Die Frankfurter Schule* (Munich: Hanser).
Wischke, M., 1994: *Die Geburt der Ethik. Schopenhauer - Nietzsche - Adorno* (Berlin: Akademie Verlag).
Witkin, R., 2003: *Adorno on Popular Culture* (London: Routledge).

Chapter 7
Historical Forerunners of Policy Analysis

7.1 Introduction

In the early seventeenth century, Germany turned into a provincial political conglomerate for several reasons.[1] Firstly, the *trade centres moved westward* to Holland and Britain. Meanwhile, the old commercial bourgeoisie of important centres such as Nuremberg, Augsburg, Cologne, and the Hanseatic cities near the North Sea and the Baltic, such as Lübeck, Rostock, Danzig, Hamburg and Bremen, declined. Furthermore, the Thirty Years War (1618–1648) with its *confessional split* led to a dogmatic narrowing of political thinking in Catholic as well as in Protestant areas. The 'Holy Roman Empire of the German Nation' (according to an old joke, neither Holy nor Roman) now was also no longer 'German' because it became an *object of foreign intervention* from France and Sweden. Three powerful German states, Brandenburg, Saxony and Bavaria, concluded secret treaties with France which contained a clause that these states would no longer vote for an Emperor from the House of Habsburg—a clause which according to the treaties of the Peace of Westphalia (1648) was close to 'imperial treason' (von Aretin 1993 I: 359). And last but not least, the *German language* had greater difficulties than French or English in overcoming Latin as a scientific language, though Luther had greatly contributed to creating a modern German language which used expressions close to the people's way of talking (cf. von Beyme 2009).

All this contributed to Germany's turning into a conglomerate of political provinces. The Empire being in such a chaotic political situation, but few German thinkers were internationally recognized, examples being Johannes Althusius and Samuel von Pufendorf. The dominant influences in Germany came from the Dutch intellectual Justus Lipsius (1547–1606) and from French and British philosophers. Nevertheless, there was an intensive political and legal debate in many of the

[1] This text was first presented at the University of Münster in Germany as "Historical Forerunners in Policy Studies" and has not previously been published.

K. von Beyme, *Klaus von Beyme*, SpringerBriefs on Pioneers in Science and Practice 14, DOI: 10.1007/978-3-319-01535-4_7, © The Author(s) 2014

newly created universities, starting with Heidelberg in 1386, since each territory tried to instruct its administrators in a modern scientific way in order to survive.

7.2 The Discovery of Policy and the Heritage of Aristotle

Compared to Anglo-Saxon countries, the German language had the disadvantage of being unable to express the various aspects of political life in a triad such as *polity* (the constitutional framework), *politics* (the decision-making system) and *policy* (the outcome of political decision-making). Germany followed the West European tradition with a predominance of Aristotelian thinking up to early modern times. In spite of this, Germany lagged behind in the international development of political theory. Whereas some British philosophers such as Hobbes clearly opposed Aristotelian state philosophy, the Protestant universities in Germany were still dominated by it. Here it was a welcome counterweight to the universalistic and eschatological concepts of the Empire in the work of Catholic thinkers. The Aristotelian definition of a person as a *political animal* seemed to support the independence of Protestant citizens from the power of the Church over the political territories. At the same time, it was an important instrument against the scholasticism of many Catholic thinkers.

The notion of *politie* in the work of Aristotle aimed for a good and just order. The Latin translation *politica* was mostly perceived as a subject of *political ethics*. The German adaptation *politic* was introduced for the first time in the works of Ludwig von Seckendorff in his *Teutscher Fürsten Staat* (German Princely State; 1656, 1976): He claimed to avoid the usual descriptions of a general theory of politics, but aimed at analysing the 'order in most German principalities'. Thus, Seckendorff's interest was already in policies in a modern sense of the word. The quality of government and its policies was opposed to the usual general theories on regimes (*Regiments*), the latter remaining with little empirical evidence about the decisions of princes. Since Seckendorff had served as chancellor of the German territory Saxony-Gotha, he had a deep insight into the way principalities were ruled in the German Empire. Seckendorff therefore hoped to illustrate a dimension in his *Umstände einer Policey* (Circumstances of policy) which he did not find in the works of contemporary 'writers' (he used the negative term *Sribenten*). In the seventeenth century, most analysts described the political order *(polity)* or the *policies*. *Politics* as a decisional system was rarely touched on because it seemed to be the privilege of the princes with the consent of the estates. *Politics* in that sense was rather treated in the rare Machiavellian treaties or at the time of declining absolutism and in the period of the French revolution when 'the people'—organized in their estates—gained some influence. Oddly enough, the princes tried to minimize the influence of the estates by introducing welfare policies in favour of the ordinary people. These policies were not necessarily consented to by estates, which were dominated by the representatives of the upper classes.

7.3 'Reason of State' and 'Sovereignty' as New Concepts in Political Theory

The German Empire was no normal state in a modern sense. From the French philosopher Bodin (1529/30–1596) down to the German thinker Samuel von Pufendorf (1632–1694), 'sovereignty' was used as the main feature of a state. This is why Pufendorf (1994: 199) called the Empire an 'irregular entity, similar to a monster' (*'irregulare aliquod et monstro simile')*. Most treaty writers on state and politics in Germany, however, contradicted Bodin (1583, 1962 II.6: 321ff, II.1: 303) and remained conservative enough to think that the Empire was still a monarchy and not only an aristocracy or *principate* as in Bodin's work.

The development from a *polycentric system of states* to a *monocentric absolutist monarchical state* in political theory was accompanied by an adaptation of the Italian theory of the 'reason of state' in the tradition of Machiavellism—more often in Catholic territories, but later on also in Lutheran States. It first took the shape of a variation on old theories about the *arcana imperii*, the secrets of imperial power. But it was no accident that the rationalist theory of 'reason of state' promoted by Giovanni Botero (*Della region di stato*, 1589) in the German translation published in Strasburg in 1596 was smoothed down to *Gründlicher Bericht von Anordnung guter Policeyen und Regiments* (Fundamental report on the arrangement of good policies and political regiments). 'Regiment' was used for the Anglo-Saxon equivalent of *polity*. The meaning of the expression *policy* included the outcome of monarchical rule in terms of decisions. The Jesuit Giovanni Botero (1544–1617) dedicated his influential book to the Archbishop of Salzburg, Wolf-Dietrich von Raitenau, who, however, did not govern according to the essence of this book and was one of the few German princes who finally lost his rule because he utterly exploited his citizens, by imposing enormous expenses for the construction of palaces and churches. Botero's work represented a kind of modern *Welfare Machiavellism*. Due to the devastation of the country in the 30 Years War, Botero did not stick to Machiavelli's main interest in internal security, defence and foreign policy. He almost invented new policy areas, such as population policy and social welfare policies, in order to win back the population's trust in their princes—princes who had ruined their state fortune with continuous wars.

The early Italian Machiavellians Botero and Boccalini had already differentiated between 'good and bad reason of state'. Their German followers concentrated predominantly on the good side of the 'reason of state'. Botero's treatise was frequently published in one volume together with his essay *About the Causes of the Grandezza and Magnificence of the Cities* (1588), which dealt with demographic policies. Botero therefore was sometimes considered as a forerunner of Thomas Robert Malthus (1766–1834), the famous British economist and demographer. The growth of the population was considered state capital, but only on the condition that it did not cause unemployment and poverty. 'Workforce not capital' was considered the foundation of wealth in a state. Infrastructure as a requirement for good trade conditions was emphasized. Botero and his German followers avoided the exaggerations of the utopian Italian literature on cities that regulated all the details—even visits to prostitutes. 'Reason of state' as a leading policy device was mainly adopted in Catholic German territories, whereas *sovereignty* was emphasized among those Protestant political thinkers who opposed the Catholic emperor in Vienna, such as Samuel Pufendorf, Hermann Conring and Bogeslav von Chemnitz. Germany was penetrated by the influence of foreign powers and it was not by chance that the intellectual Conring received subventions from the King of France and that Pufendorf and Chemnitz were for some time in the service of the Lutheran kings of Sweden, who beegan a kind of imperialism south of the Baltic on German territory (von Beyme 2011).

It was not until the eighteenth century that these denominational differences in perception of the idea of a 'reason of state' withered away, when Fichte (1762–1814) and Hegel (1770–1831) adapted this notion in their work. Fichte's works on the 'reason of state' included a foreign policy dimension, as they were occasionally directed against Napoleon's rule in the declining Empire, which was abolished under French pressure in 1806.

7.4 The Special German Development of a Polizei-Staat (State of Police)

The term 'policy' or *Policey* also evolved in its own way: some time later, some German princes narrowed down the term 'policy' to 'police' (*Polizei* in German). The German term for policy was *Policey*, which was increasingly absorbed by state activities in the sphere of internal security. The second connotation of 'policy' was lost—one of the reasons why Germany had to adapt the English term 'policy' in modern discussions (regarding the terms used for 'policy analysis' in Germany today, see Blum/Schubert in Chap. 1). Policy (*Polizei*) was reduced to a strict regulation of all social conditions of the citizens (Maier 1966: 309). It was only when civil liberties were increasingly demanded that the absolutist theory of the *purposes of the state* and the normative speculations about *happiness* were criticized (Stolleis 1992: 246). The 'policy-state' or *Polizei-Staat* grew into a term of insult in the liberal German tradition. Even the British traditions of Jeremy Bentham 'in search of happiness' have been rejected in liberal German political thought since Kant.

7.5 Development and Traditions of Polizeiwissenschaft

Modern absolutist rule increasingly differentiated the activities of the state. This caused the development of an *allgemeine Staatslehre* (general theory of the state). Later this new discipline was influential in the United States via the work of the German immigrant Francis Lieber (1800–1872), who is considered to be the founding father of American Political Science. The barren discussions about *natural law* were complemented by a positive *legal and political science* (Willoweit 1975: 364). The traditional theory of political wisdom abandoned its normative roots. However, only parts of the new differentiated branches of political thought, such as political economy and *Policey-Wissenschaft*, could match the achievements of West European political theories (Mohl 1855, 1960, I: 34).

The mainstream of the writings on the Empire remained closer to thinking in terms of a *Ständestaat* (state of estates) than to Machiavelli's doctrines on the *Principe* or to theories of sovereignty in the tradition of Bodin. Most writers on politics wrote on *policy* (*Policey*) which embraced all domestic affairs. Since the numerous territories in the German Empire were sovereign merely in their own perception, foreign policy played only a minor role, except in the biggest states such as Austria, Brandenburg (Prussia), Bavaria or Saxony. Politics was reduced to *good order and Polizey*, with increasing regulation of the social life of citizens. The theoretical concern for welfare was the positive side of this activist and interventionist concept of the state. In France the theorists of mercantilism hoped to increase the income of the state budget by state activities in production and commerce. The German doctrine of 'cameralism'—developed as a doctrine orientated towards administration and economy—envisaged more modest aims. The cameralism of the seventeenth and eighteenth centuries may be portrayed as a specific Austrian-German version of mercantilism (von Beyme 2009: 27). Some universities started as *Verwaltungshochschulen* (High Schools of Administration), for example Kaiserslautern, which was only later united with the University of Heidelberg. Since the princes needed qualified administrators, professors of 'cameralism' were frequently substituted for the older teaching jobs in *politics*, mainly dedicated to political ethics. *Policey* or *policy* was a domain of lawyers and economists, and was taught in a rather philosophical way.

New universities were founded by small Protestant states, such as Marburg (1527), Königsberg (1544), Jena (1558), Helmstedt (1574), Altdorf (1578/1622), Herborn (1584), Giessen (1607), Rinteln (1620), Strasbourg (1538/1621), Duisburg (1655) and Kiel (1665). These small state universities, for example Herborn and Helmstedt, produced leading political thinkers for a while, such as the philosophers Johannes Althusius (1563–1638) and Hermann Conring (1606–1681), who were both strongly influenced by Dutch thinkers. In the established Catholic universities, *politics* and *policy* remained a marginal discipline. All the professors in these mostly small universities wrote enormous quantities of treatises on *politics* and *policy* which remained mediocre and repetitive. One leading German historian (Meinecke 1967: 78) was shocked by the 'true catacombs of the forgotten literature of mediocrities', which he discovered in his history on the 'reason of state'. It was a period when German political philosophy in British and American histories of political thought gave short shrift to the Lutheran thinkers Pufendorf and Althusius (cf. Skinner 1978, vol 2: 341–348). This early crisis of German policy science could only be overcome when a more general philosophy of state and law was rediscovered in the leading Prussian university of Halle (with thinkers such as Wolff and Thomasius).

Lutheran thinkers emphasized 'étatism' or 'statism'. These concepts were based on the best-known verse of *Epistle to the Romans 13*, 'Render unto Caesar the things that are Caesar's'. This attitude has promoted 'a *close relationship between throne and altar*'. Philipp Melanchthon (1497–1560), one of the leading thinkers of Lutheran Church Reform, still could not find a political theory in the Bible. Therefore, those who followed him stuck to the teachings of Aristotle. Only a few thinkers, such as Dietrich Reinkingk (1656, dedication, unpaged) tried to develop a *biblische Policey* (biblical policy) in order to find a 'pleasant and fruitful policy (*Policey*) which pleases God' and that enables people 'to find rules to teach a social life which pleases God'.

In the tradition of Pufendorf, German writers combined the theory of state sovereignty with an idea of welfare for the citizens. 'Cold Machiavellian reason of state' was mostly not accepted in petty bourgeois Germany. Even a German liberal of the nineteenth century such as Robert von Mohl (1960, III) resented the fact that Machiavelli (supposedly) taught that citizens should be left to their private existence. He criticized Machiavelli for having no idea about the state's purposes and the well-being of the people. The Catholic Botero, who was a devoted follower of Machiavelli, first used the *ragion di stato* in the title of the German edition of his seminal book (1596), thereby smoothing down 'reason of state' to the German taste. The editor of this book, Lazarus Zetzner, hoped that princes would fight against social poverty and proclaim general provisions for the citizens, mostly subsumed in the term *Daseinsvorsorge* (public service tasks) in conservative thought until now.

Pufendorf was not the deepest thinker of his time, but certainly an all-round writer with some influence on Rousseau and the American founding fathers. His books on *policy* were comparatively easy to read whereas the older generation of writers on policies contributed unreadable encyclopaedias with thousands of footnotes, for example, Oldendorp, Reinkingk, and Seckendorff. From the latter, only Seckendorff (1665) structured the policies into a general concept around the notions of law, peace and welfare. Church policies and educational policy were added. The liberal jurist Mohl (1858, 1962, III) criticized the German development towards a peculiar bureaucracy whereas Britain developed administration in many fields without metaphysical super-evaluations. Britain lacked the prevailing influences of the *Junkertum* (the landed aristocracy) and the Catholic *Ultramontanism* waiting for instructions from Rome. German metaphysics of the state had only few followers in Britain, such as Bosanquet, Green and Austin.

7.6 The Decline of the Tradition of Polizeiwissenschaft

In the nineteenth century the tradition of *Polizeiwissenschaft* (policy science in the old-fashioned German sense) came to an end. Mohl was the last to write a book on *Polizey-Wissenschaften nach den Grundsätzen des Rechtsstaats* (Policy sciences according to the rules of a legal state). This book contained the two basic notions which were to follow each other in the German tradition, which changed from the absolutist 'police state' to the constitutional *Rechtsstaat*. The legal state was important because it allowed constitutional government with representation in the estates which shared legislation with the monarch. Until 1918 governments in Germany and Austria were not dependent on the confidence votes of a dominant parliament, but they had to rule in an unarbitrary way and to implement the laws of the legislative bodies. German liberal thought from Robert von Mohl down to Max Weber still preserved some elements of a bureaucratic state in the limits of the legal state.

Thinkers of German idealism, somewhat independent from political ideologies of the time, still preserved many ideas of the older 'cameralism' and sketched a rather authoritarian policy state. One example is Fichte and his work *Der geschlossene Handelsstaat* (The closed commercial state). In spite of its conservative image, Hegel's theory of the police was rather progressive in social terms: In his *Legal Philosophy* (1819/20, Sect. 236), Hegel claimed that the 'control of the police… should also take care of street lights, construction of bridges and taxation of daily needs and of the health of the people'. For Hegel, poverty (Sect. 244) was not yet creating a 'proletariat' (he used the German expression *Pöbel*, a negative modification of the French word 'people') as long as the state cared for the poor. Kant was the most explicit in refusing an enlightened despotism even if it was embellished by Bentham's *search for happiness* (see above). Kant (1964, vol 6: 158f), in polemical dispute with Achenwall, was afraid that the empty formula of happiness might lead to situations in which the monarch who wants to make his citizens happy turns to despotism. The concept might also lead to the people turning to rebellion, because they would insist on their own concept of happiness.

The most liberal version of a restriction of the state's activities was written by Wilhelm von Humboldt (1792, but not published until 1851 in Breslau, because the author was afraid of possible repression). His work, which had a great influence on John Stuart Mill, restricted state activities to external and internal security. Even educational *policy*—for which Humboldt was temporarily responsible in Prussia—was not accepted by him (1960: 69ff) because it might cause too little or too much of a result and might in the end be rather ineffective or even detrimental. The liberal Karl von Rotteck (Aretin and Rotteck 1839) wrote a polemic against the concept of a 'welfare police', stressing that the main purpose of a state was to act as a legal state. German liberals were disappointed with the authoritarian development of French radical liberals in their welfare committee. For a while, this contributed to the discrediting of welfare policies among many German liberals.

Political theory began to become liberalized in many continental countries via a theory of legislation. Pioneers of this field were the Italian thinker Gaetano Filangieri (1799, 1833: 9) who wrote a kind of encyclopaedia, and Jeremy Bentham. The British philosopher Bentham (1789, 1961: 323) aimed at a theory 'most conducive to the happiness of the whole community, by means of motives to be applied by the legislator'. Even in the old policy sciences, legislation was frequently mentioned. It was considered, however, as an *art* and not as a *science*.

Even Bentham occasionally referred to legislation as an art. Legislation in the period of absolutism was part of the *arcana imperii* (secrets of rule) of the princes and their bureaucrats (in German mostly called *Amtmänner*, office men). It was a novelty in the era of the French Revolution that the science of legislation was no longer directed only at princes and bureaucrats, but at parliamentarians and politicians.

Not until the second half of the nineteenth century did the *law of administration* lose its political policy impetus under the influence of *legal positivism*. Theories of legislation increasingly lost themselves in procedural techniques of parliamentary behaviour. Only certain conservative pioneers of administrative law, such as Ernst Forsthoff (1902–1974), combined a fairly undemocratic concept of administrative law with an important impetus towards welfare policy under the slogan of *Daseinsfürsorge* (public service tasks). This doctrine followed the older German tradition and yet it tended to be extremely modern in an un-liberal way. Not until after the Second World War was a reintegration of systematic democratic ideas of legal order and the analysis of important policies attempted (Schmidt-Aßmann 1998).

7.7 Conclusions

Protestant ethics in Prussia fused in German history with the policy ideas of an enlightened welfare state. Prussian enlightenment in the University of Halle began to subordinate theological thinking to modern rationalism. There was much loose talk about a *Prussian style*. In the field of policy sciences, this style of thinking was welfare-oriented rather than liberal and 'capitalist-minded' (Lepsius 1996). Germany was no social unit and the conservative Hegelian Lorenz von Stein (1852, 1961: 36ff) in his pamphlet on the *Question of a Prussian Constitution* argued against a constitution: for him, a constitution only made sense when the territories were basically unified in social terms. Liberals could not accept this idea because in the long run the legal state and the constitution helped to create a certain social unity in many modern states—even in Prussia, which then extended from Aachen on the Belgian border to Tilsit near the Russian border.

The German Federal Republic in its Basic Law accepted the formula of a *Sozialer Rechtsstaat* (social welfare state based on law) which abandoned the authoritarian connotations of welfare policies in the older German tradition. In the second Empire, one of the leading professors of 'law of the state', Paul Laband (1838–1918), still wrote polemics against this kind of idea because it would "drag the sublime *Rechtsstaat* into the dirt of petty social quarrels".

In the Federal Republic of Germany, the Basic Law contains a last element of the old Prussian ideas as it gives the federal legislation the right to create 'the unity of living conditions'. Oddly enough, after the reunification of two rather different Germanys this formula had to be smoothed down to "equivalent conditions of life" (Basic Law, Art. 72, 2) in order to sound more realistic.

Some political scientists have made the criticism that, with its tradition in the obedient *Policey-Wissenschaft*, modern policy science may also be in danger of applying technocratic perspectives, losing critical distancing from its objects of investigation, and losing sight of concepts such as 'power' and 'interests'. Against this background, a nuanced understanding of the still visible historical traditions of German policy analysis is essential for discussion of today's policy analysis in Germany and its advantages and disadvantages. This chapter has contributed to such an understanding by showing how 'policy' in a sense (more or less) closely related to the modern understanding of the term has developed, from its ascent in the seventeenth and eighteenth centuries to its decline in the nineteenth century.

References

Aretin, J. Ch. A. M. von; Rotteck, C. von, ^{2}1839: *Staatsrecht der constitutionellen Monarchie*, Vol. 2 (Leipzig).

Aretin, K. O. von, 1993: "Das Reich als Staat im Spiegel der Reichspublizistik" in: *Das Alte Reich 1648–1806*. Vol. I: *Föderalistische oder hierarchische Ordnung 1648–1684* (Stuttgart: Klett-Cotta): 346–362.

Bentham, Jeremy, 1789: *The Principles of Morals and Legislation* (New York) (reprint 1961).

Beyme, Klaus von, 2009: *Geschichte der politischen Theorien in Deutschland 1300–2000* (Wiesbaden: VS Verlag für Sozialwissenschaften).

Beyme, Klaus von, 2011: "Schwedisches Imperium im Deutschen Reich. Ein vergessenes Kapitel der Imperien- und Mythenbildung", in: Bluhm, H. et al. (Eds.): *Ideenpolitik* (Berlin: Akademie Verlag): 71–88.

Bodin, Jean, 1583 : *Les Six Livres de la République* (Paris: Aalen, Scientia) (reprint, 1961).

Botero, Giovanni, 1596: *Della Ragion di stato* (Turin: Unione tipografica) (reprint 1948) Appendix ((German: *Gründlicher Bericht der Anordnung guter Policeyen und Regiments* (Strassburg: Lazarus Zethner)).

Bull, H. P., 1977: *Die Staatsaufgaben nach dem Grundgesetz* (Kronberg: Athenäum).

Filangieri, Gaetano, 1799: *La scienza della legislazione* (reprint, Catania, 1833).

Humboldt, Wilhelm von, 1960: *Werke in fünf Bänden*, Vol. 1 (Darmstadt: Wissenschaftliche Buchgesellschaft).

Kant, Immanuel, 1964: "Über den Gemeinspruch", in: W. Weischedel W. (Hrsg.): *Immanuel Kant: Werke in sechs Bänden*, Vol. 6 (Darmstadt: Wissenschaftliche Buchgesellschaft): 125–172.

Knemeyer, F.-L., 1979: "Polizei, in: Brunner", in: Otto et al. (Eds.): *Geschichtliche Grundbegriffe*, Vol. 4 (Stuttgart: Klett): 875–897.

Lepsius, M. Rainer, 1996: "Die pietistische Ethik und der Geist des Wohlfahrstaats oder: Der Hallesche Pietismus und die Entstehung des Preußentums, in: Clausen, L. (Ed.): *Gesellschaften im Umbruch* (Frankfurt: Campus): 110–124.

Maier, Han, 1966, ^{2}1980: *Die ältere deutsche Staats- und Verwaltungslehre* (Neuwied: Luchterhand; Munich: Beck).

Meinecke, Friedrich, ^{3}1963: *Die Idee der Staatsräson in der neueren Geschichte* (Munich: Oldenbourg).

Mohl, Robert, 1858: *Die Geschichte und Literatur der Staatswissenschaften in Monographien dargestellt*, Vol. 3 (Erlangen: Enke) (Reprint: Graz, Akademische Druck und Verlagsanstalt, 1960).

Pufendorf, Samuel von, 1994: *Die Verfassung des Deutschen Reiches (De statu imperii Germanici)*, bilingual edition (Frankfurt: Campus).

Reinkingk, Dietrich, 1651, ^{5}1653, 1701: in: Zubrodt J. P. (Ed.): *Biblische Policey* (Frankfurt: Bencard).

Schmidt-Aßmann, Eberhard, 1998: *Das allgemeine Verwaltungsrecht als Ordnungssidee* (Berlin: Duncker & Humblot).

Seckendorff, Ludwig. Von, 1665: *Teutscher Fürsten Stat*, Vol. 2 (Frankfurt: Glashütten Auvermann) (reprint 1976).

Skinner, Quentin, 1978: *The Foundations of Modern Political Thought*, Vol. 2 (Cambridge: Cambridge University Press).

Stein. Lorenz von, 1852, ^{2}1961: *Zur preußischen Verfassungsfrage* (Darmstadt: Wissenschaftliche Buchgesellschaft).

Stolleis, Michael, 1988: *Geschichte des Öffentlichen Rechts in Deutschland*, Vol. 1. *Reichspublizistik und Policeywissenschaft* (Munich: Beck).

Willoweit, D., 1975: *Rechtsgrundlagen der Territorialgewalt* (Cologne: Böhlau).

Wolzendorff, K., 1918: *Der Polizeigedanke des modernen Staates* (Breslau: Korn).

Chapter 8
Political Institutions: Old and New

'Political institutions—old and new' as a topic has two dimensions: the evolution of old and new institutions and the reflection of these developments in political theory. There is, however, an asymmetry of these dimensions.[1]

8.1 From Old to New Institutions

Few really 'new institutions' developed in the three waves of democratisation after 1789. The three major branches of public life existed not only in Montesquieu's theory. But their weight had shifted, especially in tune with the decline of *monarchical power*. The first old institution which spread all over the world—with the exception of the United Kingdom—was the 'constitution', mostly considered as an emanation of the popular will, and since 1918 frequently submitted for ratification by a popular referendum. The revolutionary constitutions in France (since 1792) and in the United States (in 1787) did not completely break with the institutions of the pre-Revolutionary regime, but adapted them to the needs of representative—and later when universal suffrage was accepted—democratic government. Constitutions by the conservatives of the early nineteenth century were considered as 'revolutionary institutions'. But under the threat of Revolution various forms of adaptation of this institution by the existing monarchies took place. Constitutions were either imposed by monarchs (octroi), as the Piedmontese 'Statuto Albertino' of 1849 which was to become the constitution of the kingdom of Italy, or negotiated by legislatures and monarchs (France 1792, Spain 1810 and in many European territories after 1815). Even dictatorships normally adapted some kind of constitution, including a bill of rights which the regime rarely respected.

[1] This text was first published by Klaus von Beyme as "Political Institutions—Old and New". In: A. W. Rhodes et al. eds: *The Oxford Handbook of Political Institutions*. Oxford, Oxford University Press, 2006: 743–757, 2nd edition, 2008. By permission of Oxford University Press (www.oup.com) that was granted on 13 May 2013 by Ms. Mary Bergin-Cartwright.

K. von Beyme, *Klaus von Beyme*, SpringerBriefs on Pioneers in Science and Practice 14,
DOI: 10.1007/978-3-319-01535-4_8, © The Author(s) 2014

Old assemblies of 'estates' were transformed into *modern parliaments*, sometimes as late as 1866 in Sweden. Various forms of advisers to the crown developed into a *modern cabinet* with a *prime minister*. Important institutional changes were grounded not so much in the internal change of institutions, but in their mutual relationship within the system. The major institutional innovation was the development of dependence of cabinets on the confidence of parliamentary majorities over almost one century. It happened in systems with continuity of former estate systems (Britain, final conflict 1832, Netherlands 1868, Sweden 1917). New institutions were created by new revolutionary systems which established *parliamentary responsibility of governments* (France 1830 and Belgium 1831). Parliamentarisation of neo-absolutist regimes was normally late—with the exception of Italy (1860). The latest latecomers in this group were Germany and Austria (1918). Parliamentarisation did not evolve in harmony with the extension of voting rights. *Suffrage* in the first parliamentary systems on the continent was hardly above 1–2 %. Germany introduced universal suffrage as early as 1871, but full parliamentary responsibility of governments followed only in 1918 (von Beyme 2000: 28).

Most regimes in the nineteenth century were dualistic constitutional monarchies. Revolutions which led to a Republican system—as in France in 1848, in Germany and Finland in 1918—tried to find a Republican equivalent for a system with a president elected by popular vote and not depending on parliamentary majorities. Only in the Fifth French Republic this type of government was dubbed *semi-presidential regime*. Frequently it evolved in a constitution-making process with extensive debates on the virtues of the American 'presidential system'. Finally a European compromise led to a hybrid of parliamentary systems in which the prime minister and the cabinet depended on parliamentary votes and the president was equipped with the right to dissolve parliament as a counterweight against permanently hostile legislative majorities (von Beyme 1987: 33ff).

Two major institutions had existed already in Ancient Rome but developed into powerful organisations which penetrated the whole life of society: bureaucracy and parties. *Bureaucracy* for Max Weber was the dominant institution of modernisation. *Parties*—frequently discriminated as unpleasant extra-constitutional and anomic institutions under the label of 'factions'—only in modern times became the basic element which coordinated all the institutions of the state.

An exception to the 'nothing-new-under-the-sun approach' to institutions was the success story of *constitutional courts*. This institution was new only if we exclude the functional equivalent of the American Supreme Court which developed—not completely in tune with the ideas of most founding-fathers of the Constitution—judicial review of legislative acts since its seminal decision Marbury versus Madison in 1803. In the light of former colonial history the USA did not accept special courts because the American states were afraid of a continuation of the 'Star chamber proceedings' of the British Crown. Not even a special constitutional court was feasible. Therefore, the drafters of the American Constitution deliberately did not accept *abstract judicial review*. The Supreme Court was the least democratic decision-making body and it was meant by the Federalists to serve—as the Senate—as another check on volatile democratic decisions in an elitist deliberating institution with no direct access of the people.

It is an exaggeration that judicial review after 1945 was accepted "at the point of a gun" (Martin Shapiro). Only Japan followed the American model. In Europe the 'Austrian model' was accepted, developed by Hans Kelsen in 1920/21. Kelsen (1922: 55) was inspired by the 'Imperial Court' of the 'German Confederation' ant its revolutionary Constitution of 1849 which envisaged already the 'constitutional complaint' (Sect. 126 f, g). This type of judicial review became prominent in the European model, which largely followed the German example. A variation of a constitutional court sprang up even in political cultures such as France in the 'conseil constitutionnel'—a country which originally was hostile to the very idea of 'judicial review' against laws and acts of 'the state' because it contradicted the French Republican tradition of popular sovereignty.

Some institutions spread from one area to others, such as the *ombudsman*. This office was not really new. Ombudsmen were even remainders of pre-democratic enlightened absolute rule as a safety valve for individual complaints. New institutions such as *planning authorities* were developed in an era of a rational optimism that society can be shaped by the state. But they withered away in the wave of neo-liberalism which followed the collapse of communist systems and the high days of the welfare state. New institutions with a political impact were also developed to guarantee a balance between the economic institutions. A *national bank* and *committees for the control over monopolies* gained influence. The market system no longer looked for democratic socialist institutional schemes but tried indirectly to steer the economy by independent institutions.

Institutional theories always developed in cycles after revolutions (1789, 1830, 1848, 1871, 1918, 1945). Never did so many regimes break down at one time as in 1989. Never were so many regimes transformed from one fairly uniform Communist institutional type to another fairly uniform type of Western democracies. 'Special national roads of development to democracy' were no longer hailed, as in the period after the first World War. At no time did so many countries launch an institutional debate as in the 'new democracies'. 'Constitutional engineering' became the highly misleading basic term of the new branch of *transitology* in the third wave of democratization (Sartori 1994). 'Grandpa's institutionalist Political Science' was again en vogue. Old-fashioned debates on the preference of semi-presidentialism versus parliamentary systems were revitalized. Old institutions such as the one-party monopoly, the collective presidium of the legislature as an equivalent to Western presidents, planning offices, the wide range of competences of a *prokuratura* which was more than a prosecuting attorney, and the gigantic bureaucracies of state security had to disappear. The new institutions, however, were the old ones—mostly institutions from Western countries. Most influential institutions proved to be the French semi-presidential system and the German Constitutional Court. Many details of institutions were copied from a five-percent-threshold for parties during elections to electoral laws, votes of constructive non-confidence, or abstract judicial review (von Beyme 1996: 98f).

8.2 The Evolution of Theories of Institutions

8.2.1 Theories and Methodological Approaches to Institutions

Theories tended to be changing more quickly than the institutions they had pretended to analyse. Quite frequently theories of institutions lagged behind the real functioning of a system, such as Montesquieu's doctrine which ignored the institutions of parties and adhered a schematic view of the British system. Some older theories of politics started from the assumption that Political Science as a whole works with an institutional approach, whereas sociology emphasizes the aspect of stratification (Allardt 1969: 17). This assumption was never correct. Even older approaches combined 'elite'—a more important notion in American social sciences than 'class'—predominant in European sociology with institutional studies. This concept neglected the necessary differentiation between theory and method. Elites or stratification are basic notions of social theory. The institutional approach, on the other hand, belongs to the methods of political science. A theory can be falsified. Methods, however, survive even if certain theories which have been applied with the help of certain methods proved to be wrong. The institutional approach is not obsolete when the old institutional paradigm of a 'separation of powers' was no longer applicable to modern parliamentary systems. Older institutional theories amalgamated elements of new theories such as the theory of pluralism and methods which went beyond the old-fashioned juridical normative approach to institutions. Theoretical concepts like 'pluralism' or 'federalism' can be put into empirical operation with institutionalist, behavioralist or rational choice methods.

Political science initially tried to legitimize itself by a revival to the Aristotelian concept of politics. It tended to favour the institutional approach—compared to eschatological theories of politics from St. Augustin to Marx. The virtue of the classical institutional approach was that it started from the assumption that the political process is 'open' in principle, and full citizens are basically 'equal'. No ontological essentialist differences between princes and the people, enlightened elites and humble subjects, or proletarian avant-gardes and the masses were accepted.

Classical institutionalists from Montesquieu to Tocqueville were never naïve ontological analysts but described institutions in comprehensive social settings of a system. Each institution was linked to a special promoting social group. Only rarely were deistic or mechanistic metaphors of a clockwork applied in a formalistic way to political institutions. The 'mechanics' of institutions included contradictory elements, such as in inter- and intra-institutional conflicts in two-chamber systems of parliaments and the difference of government and opposition. Most institutional theories favoured a procedural concept of politics. For Max Weber the typical occidental development—deviating from the rest of the world—can be explained by institutional differentiation of religious and secular power. The most interesting institution for Max Weber was the constitution of 'cities' which were not mere agglomerations around a power center and which deviated from the pattern of patrimonial and feudal systems of rule ('Herrschaft'). From Max Weber to Stein Rokkan 'modernisation' in politics was basically understood as a process of institution-building. Contrary to economic modernisation theories institutions such as bureaucracies or the military were seen as the momentum of modernisation. Weberian concepts were influential: bureaucracies were superior to parochial or feudal elites. Beyond Weber some analysts preferred bureaucratized party politics to bureaucratic rule.

A social concept of institutions gradually differed from merely normative legal and political theories. In Britain Barker (1961: 166) suspected even after 1945 that most institutionalists hailed their preferred institution as a disguise for a cult around a social group. In the French legal theory of law Hauriou (1906) tried to avoid this danger of the old institutionalism by the differentiation of 'institution-chose' having objective dignity and the *institution-groupe,* suspected of being only disguised selfish group interest. Group theories of institutions were mostly unable to agree on the relative weight of certain institutions. The continental Roman law tradition suggested that 'the state' was the most important institution, whereas a leftist British tradition from guild-socialism to Harold Laski insisted that the state was just another 'collective group'. Anglo-Saxon theories—with the exception of some Hegelians in Britain from Thomas H. Green to Bernard Bosanquet—nourished a deep distrust against the notion of the state and rather preferred 'government' as the central notion for institutional analysis.

The development of institutional theory after 1945 proved to be oscillating between waves of neglect and rediscovery of institutions. The attempt to make political science finally scientific stood against the accepting institutional analysis as the centre of research. The *new science of politics* in the USA used the term 'institution' in the vague sense of neighbouring social sciences, such as sociology or anthropology, as "a pattern composed of culture traits specialized to the shaping and distribution of a particular value (or set of values)" (Lasswell/Kaplan 1950: 47). The *behavioral revolt* was directed against the old institutionalism, but did not avoid institutions altogether. Eulau (1969: 1, 158), a pioneer of the 'behavioural persuasion', developed a synthesis of 'behavioral-institutional research', mainly concentrated in legislative and judicial studies. Whereas Eulau critically worked on a theory of micro–macro-relations—in spite of the basic individualism of this approach—later behavioralists frequently uncritically generalized the findings on the micro-level in macro-politics of the institutions. The 'epitaph of a successful protest' which Dahl (1969) proclaimed in 1969 was premature in the eyes of later analysts. Wahlke (1979) in his presidential address for the American Political Science Association 10 years later was more sceptical. After a quantitative analysis of review articles and research notes in the *American Political Science Review* he came to the conclusion that old-fashioned institutional studies prevailed even in this Review which was considered to be the 'battle organ' of the victorious behavioral revolt.

Behavioralism was accused of lacking theory-building. *Systems theory* hoped to heal this shortcoming. System's theories in America had the virtue to develop—for the first time since Weber—a generalized theory of institution, overcoming the short-comings of ad-hoc-theories in Europe. For Parsons, deeply influenced by Weber, institutional patterns, perceived in a demystified way, were the backbone of social systems. Only in later variations of the theory of systems did ‘structures’ become more important than institutions. They had, however, no predetermined role. Similar functions within the system were completed by very different structures. The early Luhmann, originally Parsons’ devoted disciple but soon a defector who created his own autopoietic version of a theory of systems, still used institutions and structures as synonym: “Institutions are behavioural expectations generalized in temporal and social dimensions, and thus create the structure of social systems.” System’s theory created a new methodological terminology, but on the descriptive level it classified the traditional powers, such as the executive and parliament, adding bureaucracy and parties. They got, however, more scientific names such as ‘rule-setting’, ‘rule applying’, ‘rule adjudicating’ and ‘rule-enforcing’ institutions.

Institutions in the new approaches such as behavioralism or functionalism were no longer independent entities and were dealt with—according to research questions as ‘independent or dependent variables’—just as other elements of analysis. In ‘structural functionalism’ the systemic needs of the social system tended to produce political institutions needed to solve the basic problems any society (Eisenstadt 1965). Thus, the analysis ended in a global justification of all the institutions developed in various societies. ‘Historical institutionalism’ was closest to treat institutions, such as ‘the state’, as the independent variable. The impact of institutions was studied over time—from the way political groups defined their interests to policy-outcomes under various regimes (Steinmo/Longstreth 1992). The old generalisation of modernisation theories was overcome. Researchers discovered the dependence of policy-outcome on historical institutions and decisions which could not easily be changed by political actors. Policy results proved to be ‘path dependent’. A variety of models—particularly in the field of welfare policies—was discovered (Esping-Andersen 1990). The new institutionalism can better account for the paths that political actors will follow in order to arrive at the prescribed equilibria.

Behavioralism and functionalism were the major foes of the old institutional school represented by Carl J. Friedrich or Herman Finer. The old institutionalism paradoxically got theoretical support by radical political thinkers who opposed the institutions of the existing Western democracies. *Neo-Marxism* and *radical post-behaviouralist approaches* brought the ‘State back in’ even in the American discussions. But political institutions were always the dependent variable, the independent variable was the economic subsystem of society. System’s theory reacted in a hostile way to the new debate on the state. Easton (1981: 322), a pioneer in substituting the ‘political system’ to old-fashioned theories of ‘the state’, was afraid that the neo-radical wave in political theory—mostly developed in Europe from Miliband to Poulantzas—might end up in a ‘romantic backlash’ and that the state would start to besiege the political system. Easton’s misgivings were exaggerated. Neither Neo-Marxism nor Neo-Conservatism elaborated a new metaphysical concept of the state. But since these new approaches concentrated on the economic aspects of the relationship of state and society they failed to develop a differentiated theory of institutions. At the end of the neo-radical movement which had influenced the development of political theories the holistic theories of the 1960s were approaching each other.

The new wave of the *policy-approach* in the 1970s ended in a merger of systems theory and neo marxist state theories. A central actor was needed and though many empirical scholars no longer called it 'the state', a great variety of actors and their institutions were introduced in order to demonstrate the genesis of a decision—or of a 'non-decision'. *Network approaches* discovered so many veto players to avoid the impression that one actor, such as 'the state' was still considered as an ontological entity as in some older institutionalist theories (Tsebelis 2002).

The *rational choice school* offered another approach which rediscovered the institutions. The bias of this school was that theory perceived social systems as consisting of only utility-maximising rational individuals. They engage in strategic interactions which stabilize an equilibrium. This approach was highly quantifiable but its predictive capacities were rather limited because apparently non-rational collective and ideological motives distorted the 'necessary outcome' of the prognosis. Political institutions—such as parliamentary groups and their leaders—had to explain why the 'normal behavior' within larger institutions, such as parliaments, did not function in the utility-maximizing way the strict individualism of the theory had envisaged. The rational choice approach had the virtue of making cooperation in institutions plausible as far as norms of cooperation were internalized. These norms, however, hardly rise with one institution. They are pre-existing to most institutions, and only historical political culture studies can enlighten us about their genesis. Social institutions apparently determined policy outcome and even economic performance of systems. *Organisational theory* discovered these institutions in many fields—from legislation till industrial relations (Streeck 1992).

8.2.2 National Traditions and Transnational Diffusion of Institutions and Theories About Institutions

Institutional theories developed in tune with national traditions of institutions. Continental 'statism' has always differed from Anglo-Saxon concepts which did not accept a dogmatic typologies of 'state and society'—the expression of a historical compromise between monarchy and the legislative powers of 'estates'—from Hegel to Lorenz von Stein. In spite of many typologies of the role of institutions in various political cultures, the dynamics of institutional theories was never strictly limited to national traditions. The more radical-minded constitution-makers and political theorists worked in their countries, the stronger was the influence of *foreign models.* After 1789 and after 1848 the French model had some impact on the

Continent. The French model of a so-called 'unauthentic parliamentarianism' later was less attractive than the British model for liberals in Europe. France, moreover, was constitutionally unstable. According to a famous anecdote a British traveller who asked in Paris for the French constitution got the ironical answer from the book dealer: 'Sorry, we don't carry periodical literature'. The opposite example was the American system, frequently admired for its sheer institutional stability over time. For certain parties in Europe the American model was hailed because the American system was considered as being only 'political'—not aiming at a complete change of social powers in the society as did the French revolutionary model from Hannah Arendt to Dolf Sternberger.

The theory of institutions was strong in American anthropology and developed some impact on the neighbouring social sciences. A long debate was launched between 'diffusionists' who thought that social institutions developed from one centre to other areas (Thor Heyerdal even tried to demonstrate the possibility of diffusion of institutions by imitating boat trips from Polynesia to South America) and the 'functionalists'—prevailing in America—considered the development of social institutions rather as the result of social needs which led to functional equivalents of rather similar institutions. The political debate in the North Atlantic world was, however, more diffusionist than in the realm of cultures preserving only oral traditions. 'Institutional engineering' in political systems relied on a huge bulk of constitutional models and political theories which shaped them. Conscious adaptations of foreign institutions merged with national traditions since the belief that national institutions 'grow' out of national traditions—widely accepted by conservative parties in the nineteenth century—was withering away.

The USA never shared the cult of the state as a major institution. Nevertheless the citizens were more proud of their institutions than in other countries. The Civic-Culture study by Almond/Verba (1963: 102) found out that 85 % of the Americans were proud of their institutions, but only 46 % of the British, 7 % of the Germans and 3 % of the Italians. Already one of the first European evaluations of the American system, by Lord Bryce (1988, 1959, vol. 1: 1) was puzzled by a typical American question: 'What do you think of our institutions?' which he never heard in Britain. American preference for institutions was explained by the lack of a cult of personality and monarchical symbolism.

8.2.3 Institutional Crises and the Para-theories of Institutions

Later theories had to cope with the fact that attitudes towards institutions are not permanent features of some kind of 'national character'. Periodical crises of national institutions less inspired the creation of new institutions than the development of new theories on institutions. Most of them hardly deserved the term 'theory' and were *ad hoc-generalizations* which did not survive in long-term developments. Crisis-mongering leads to much discussed best-sellers in the intellectual sphere which contributed at best para-theories. Cycles of corruption and unlawful practices can undermine the basic confidence in the institutions.

Huntington (1981: 4) found a general gap between ideal and institutions—the so-called IvI gap—as 'a peculiarly American form of cognitive dissonance.' The message was not without hope in Post-Watergate-America. Ideals of the American creed periodically purify and revitalize American institutions. In other countries another crisis of institutions was criticized. The scenarios were frequently even more pessimistic. New institutions seemed to undermine the old constitutionally guaranteed institutions. The 'new social movements' caused fear and misgivings. They may have been centred in Berkeley, Paris or Berlin, but they spread all over the world and formed loose revolutionary networks.

After the students' riots in the Western world, combined with protests against the Vietnam war, a new wave of crisis-of-institutions-theories swept over Western democracies. In Germany sociologists, such as Schelsky (1973: 21), suspected that a 'revolutionary march through the institutions' might undermine the system. No system's change happened. The only long-term consequence was that former students' rebels in 1998 entered the Federal Government. Germany, as a country of conservative institutional immobilism, all of a sudden became the 'Mecca' for a new institution, the ecological 'Green Party'.

In France the sociologist Crozier (1970) came to rather far-reaching conclusions with his fear that a society is in danger where institutions block each other and lead to non-decision. Under the temporary pressure of the students' rebellion in 1968 the historical fear sprang up again that French systems proved to be unable to reform their institutions. The traumatic inspiration from French history which dooms the country to develop by periodical revolutionary system's changes led to a premature prognosis. The French Fifth Republic survived, though de Gaulle withdrew earlier than expected, whereas the Italian system collapsed, but at a time in the early 1990s when the storms of para-revolutionary unrest had smoothed down. There was a lot of theory-building on a second Italian Republic, but the changes of the system hardly justified speaking of an institutional revolution. The party system was the only institution which was substantially affected by the institutional crisis of the system. The 'new Republic' proved to be the 'old Republic'. The syndicalist enthusiasm for new social movements without bureaucratic structures which endangered established institutions since 1968 was met by new institutional arrangements of the old institutions. 'Neo-Corporatism' in Northern Europe had to explain why regimes did not collapse in a crisis of the institutions. From 1985 to about 1995 no book on the relationship between state and society was successful unless it contained the catch-word 'Neo-Corporatism'. Ten years later no book could be sold if it still stuck to this paradigm. Neo-Corporatism has withered away under the sun of neo-liberalism. Together with the term 'ungovernability' for which it was meant by Schmitter to serve as a remedy, neo-corporatism showed another time how short-living theoretical fashions are—especially in the realm of institutions which invite more than other subjects to simplistic every-day evidence in the style of theorising.

8.3 Conclusion: Personal Academic Experiences and Prospects for Institutional Studies

Institutions develop less quickly than theories on institutions. 'Historical institutionalism' has demonstrated that institutional traditions are not easy to change. Institutions which have lost their former justification, such as certain ministries or state agencies adapt new purposes and continue to exist. Even oddities like the electoral college in the USA or an 'executive second chamber' in Germany from Bismarck to Adenauer have not been changed in spite of numerous reform initiatives. Even the occupation forces in Germany after 1945 failed in trying to impose on West Germany different systems of a federal chamber, different forms of industrial relations or a unified social security system. The cold war soon promoted other priorities than the overhauling of traditional institutions.

Organizational theory has developed many strategies for the reform of political institutions. They were more successful in the revived 'new institutional economics' in the context of enterprises and industrial relations (Richter 1994: 3). The 'new institutionalism' in political science, however, has to live the fact of the persistence of many forms of organizational routines and structures. Most institutional reform proved to be ad hoc activity (March/Olsen 1989: 69ff).

There is a permanent division in political science between the 'hard' type of analysis aiming at universal laws—as in behavioralism and rational choice—and the 'soft' historically oriented analysis of political events and lines of cultural development. The hope remains that both camps engage in a fruitful exchange (Rothstein 1996: 156). The 'Neo-Institutionalism' was a major step in the direction of this synthesis. March/Olson (1984: 747) hoped for a 'gentle confrontation between the wise and the smart' which characterises innovations in intellectual history. Many movements and theories have called themselves 'new'. As in other fields—such as art—they quickly ended in 'post'-movements. In the best case this lead to a development 'from post- to neo.' Is neo-institutionalism really new? (1) It differs from the older institutionalism in the attempt to work theory-oriented. (2) It contains the achievements of former revolts—such as the behavioural and the rational choice-revolts—to differentiate between dependent and independent variables, though some authors blur this difference and treat their institutions simultaneously as dependent and independent variable (Pedersen 1991: 131f). Neo-institutional approaches observe actual behavior instead of legal and formal aspects of political behavior which prevailed in older theories. (3) The main virtue is that concepts have been developed which make new institutionalism more comparative than the older juxtapositions of regimes in early institutionalism (Peters 1996: 206).

Comparative studies on institutions in Europe developed between European traditions and American innovations. The first foreign influences on my own thinking took place in France in the late 1950s. As a student in France Duverger and Aron have exercised considerable influence. My book on 'Political Parties' (1985) sometimes has been dubbed as an 'updated version' of Duverger's study. This perception did hardly justice to my own intentions: comparative studies of institutions according to my interests had to get rid of three vices of the older institutionalism in France: (1) The preoccupation with a unilinear causality between electoral laws and parties in the school of André Siegfried and Duverger. (2) The benign neglect for foreign languages besides French and the lacking interest in 'Smaller European Democracies'. The project under this title, developed by Stein Rokkan and Hans Daalder was seminal for my own studies on parliaments, parties, interest groups and trade unions. (3) The study of institutions without reference to policy-outcomes.

My own academic socialisation in political science was affected by American theoretical developments in two waves. As a 'true disciple' an old institutionalist, Carl J. Friedrich, I carefully followed at Harvard University the lectures of Friedrich, V. O. Key, W. Y. Elliott and McCloskey. The new developments, however, took rather place at the sociology department. Two German students in 1961/62 went to the courses of Talcott Parsons: Niklas Luhmann and myself. Only the former became a true disciple of Parsons'. Institutionalists like me felt a subversive joy of pilgrimage to MIT in order to study with Lasswell as a visiting professor and Karl W. Deutsch. The second personal involvement took place when I was a visiting professor at Stanford University and underwent influences of my colleagues, Gabriel Almond, Seymour Martin Lipset and Heinz Eulau. My own work was shaped by a moderate deviation from 'Palaeo-Intitutionalism' in a turn to sociological views in the tradition of Karl Deutsch and Martin Lipset.

In Germany 'the state' was no longer a subject for political scientists like Dolf Sternberger and Carl J. Friedrich who run the Heidelberg institute. The state after Nazi rule was considered as the incarnation of 'mislead nationalism'. Institutions were kept free from 'identity politics' which only in the age of post-modernity became a new concern of political science. Identity-building was promoted in a rational way, via 'constitutionals patriotism' in German theories from Sternberger to Habermas. 'The state' of the older German 'Staatslehre' was no longer a concern. The problem with state institutions was rather an almost silly anglophile bias in the studies on parliamentary systems and electoral laws, initiated by F. A. Hermens, D. Sternberger and others. Institutional theory was frequently dependent on political reforms. There was a period when the 'Grand Coalition' in Germany (1966–1969)—with advice from many political scientists and jurists—seriously planned to introduce the British relative majority electoral law, in the hope that only a two-party-system would survive. But even early political culture studies had a certain bias in favour of the 'Anglo-Saxon' model. With Almond's neglect of the consociational democracies which he lumped into one category of hybrids between the British and the 'continental' model, consisting of the Benelux countries and Scandinavia, the younger generation had to take issue. Arend Lijphart and Gerhard Lehmbruch—with whom I worked in an Institute at Tuebingen—have enlightened me more than the traditional state-orientation of the 'nestor' of German Political Science, Theodor Eschenburg, at that time my colleague at Tuebingen (cf. Daalder 1997: 227ff). The younger generation on the continent discovered the traditions of 'consociationalism,' divergent from British winner-takes-all concepts.

My own work differed increasingly from Carl Friedrich's in two respects. The impact of American political sociology directed my interest to elites, interest groups and trade unions (1980) which were undeveloped in European comparative studies. In studies on Communism Carl Friedrich emphasized totalitarianism with a static bias. The neglect of interest groups was also detrimental to studies on Eastern Europe. No internal conflict and development was possible. Even Friedrich's co-author, Z. Brzezinki, was no longer able to follow Friedrich and did not participate in the second revised edition of *Totalitarian Dictatorship and Autocracy* (1965). I came into a conflict of loyalty with my teacher because I was not willing to substitute Brzezinski. Since my studies in Moscow (1959/60) I was more able than the older generation of Sovietologist to discover modest steps towards liberalisation and erosion of dictatorship. Moreover, in comparative studies in both, East and West, I was interested in institutions not per se, but in combination with their impact on policies (1982). In so far I was a 'neo-institutionalist' before the label has been invented.

The most interesting institution for older institutionalists, like Friedrich, was federalism. Especially when they worked on the institutions of the budding European Community they started from the normative assumption that federalism was 'progressive' per se. Doubts of the rational choice school in the work of Riker (1964) who calculated the costs of federalism by reluctant veto groups in the decision-making process and especially in the implementation of decisions of the national level, were widely ignored in Europe. In recent studies on federalism I turned rather to comparisons of federalist and decentralized unitary states. Only in the 1990s scholars from smaller European countries, like Switzerland, Sweden or the Netherlands (D. Braun, H. Keman), discovered that decentralized non-federal states in many respects had better performances than the federalist systems. The institutional economy studies discovered in addition that the American model of a 'federalism of competition' instead of a 'federalism of joint decision-making'—does not prosper in federations with many small units and that corruption spoils the decision-making process of federal institutions.

The new wave of institutional studies in economics proved to be fertile in political science, enlarging the range of institution to many quasi-governmental institutions from the National Banks to units which administrate unemployment or protection of environment. Comparative politics as a study of institutions will certainly continue to develop in the direction of policy studies and include a greater number of actors and veto groups than recognized in the older schools of institutionalism, still largely thinking in terms of a global 'checks-and-balances' theory. Neo-institutionalism will never develop back into the old institutionalism. Even specialists on institutions who are inclined to accept their institution as an independent variable can no longer prevent that non-institutionalist approaches take institutions only as one dependent variable among others. Even a blatant nostalgia for the older institutionalism can lead only to half a comparative analysis when it excludes the other half of individual behavior of actors. Neo-institutionalism cannot substitute the behavioral and the rational choice revolts, but can only correct their theoretical and methodological exaggerations.

References

Allardt, E., 1969: Political Science and Sociology. *Scandinavian Political Studies*: 11–21.

Almond, G.; Verba, S., 1963: *The Civic Culture* (Princeton: Princeton University Press).

Barker, E., 1961: *Principles of Social and Political Theory* (Oxford: Oxford University Press).

Beyme, K. von, 1980: *Challenge to Power. Trade Unions and Industrial Relations in Capitalist Countries* (London: Sage).

Beyme, K. von, 1982: *Economics and Politics within Socialist Systems. A Comparative and Developmental Approach* (New York: Praeger).

Beyme, K. von, 1985: *Political Parties in Western Democracies* (New York: St. Martin's Press).

Beyme, K. von, 1987: *America as a Model. The Impact of American Democracy in the World* (New York: St. Martin's Press).

Beyme, K. von, 1996: *Transition to Democracy in Eastern Europe* (New York: St. Martin's Press).

Beyme, K. von, 2000: *Parliamentary Democracy: Democratization, Destabilization, Reconsolidation, 1789–1999* (Houndsmills: Macmillan).

Bicchieri, C, 1993: *Rationality and Coordination* (Cambridge: Cambridge University Press).

Bryce, J., 1888, 1959: *The American Commonwealth*, Vol. 2 (New York: Putnam).

Crozier, M., 1970: *La société bloquée* (Paris: Seuil).

Daalder, H., (Ed.), 1997: *Comparative European Politics. The Story of a Profession* (London: Pinter).

Dahl, R., 1969: "The Behavioral Approach in Political Science. Epitaph for a Monument to a Successful Protest", in: Gould, J. A.; Thursby, V. (Eds.): *Contemporary Political Thought.* New York: Holt, Rinehart & Winston: 118–136.

Dogan, M.; Pahre, R., 1990: *Creative Marginality. Innovations at the Intersections of Social Sciences* (Boulder: Westview).

Easton, D., 1981: *The Political System besieged by the State. Political Theory*: 303–325.

Eisenstadt, S. N., 1965: *Essays on Comparative Institutions* (New York: Wiley).

Esping-Andersen, G., 1990: *Three Worlds of Welfare Capitalism* (Oxford: Polity Press).

Eulau, H., 1969: *Micro-Macro Political Analysis* (Chicago: Aldine).

Göhler, G. (Ed.), 1987: *Grundfragen der Theorie politischer Institutionen* (Opladen: Westdeutscher Verlag).

Hauriou, M., 1906: *L'institution et le droit statuaire* (Paris).

Huntington, S.P., 1981: *American Politics: The Promise of Disharmony* (Cambridge: Belknap).

Kelsen, H., 1922: *Die Verfassungsgesetze der Republik Österreich*, part V (Vienna: Deuticke).

Lasswell, H. ; Kaplan, A., 1950: *Power and Society. A Framework for Political Inquiry* (New Haven: Yale University Press).

March, J. G.; Olsen, J. P., 1984: "The New Institutionalism: Organizational Factors in Political Life", in: *American Political Science Review*, 78, 2: 734–749.

March, J. G.; Olsen, J. P., 1989: *Rediscovering Institutions: The Organizational Basis of Politics* (New York: Free Press).

Ostrom, E., 1995: "New Horizons in Institutional Analysis", in: *American Political Science Review*, 89: 174–178.

Parsons, T., 1954: *Essays in Sociological Theory* (Glencoe, Ill.: Free Press).

Pedersen, O. K., 1991: "Nine Questions to a Neo-Institutional Theory in Political Science", in: *Scandinavian Political Studies*, 14: 125–148.

Peters, B. G., 1996: "Political Institutions, Old and New", in: Goodin, R. E.; Klingemann, H.-D. (Eds.): *A New Handbook of Political Science* (Oxford: Oxford University Press): 205–220.

Richter, R., 1994: *Institutionen ökonomisch Analysiert* (Tübingen: Mohr).

Riker, W. H., 1964: *Federalism: Origin, Operation, Significance* (Boston: Little, Brown).

Rothstein, B., 1996: "Political Institutions. An Overview", in: Goodin, R. E.; Klingemann, H.-D. (Eds.): *A New Handbook of Political Science* (Oxford: Oxford University Press): 133–166.

Sartori, G., 1994: *Comparative Constitutional Engineering* (Basingstoke: Macmillian).

Schelsky, H., 1973: *Systemüberwindung, Demokratisierung, Gewaltenteilung* (München: Beck).
Shepsle, K. A., 1989: "Studying Institutions: Some Lessons from a Rational Choice Approach", in: *Journal of Theoretical Politics*, 1: 131–147.
Steinmo, S.; Longstreth, F. (Eds.), 1992: *Structuring Politics: Historical Institutionalism in a Comparative Perspective* (Cambridge: Cambridge University Press).
Streeck, W., 1992: *Social Institutions and Economic Performance* (Beverly Hills: Sage).
Tsebelis, G., 2002: *Veto Players: How Political Institutions Work* (Princeton: Princeton University Press).
Wahlke, J. C., 1979: "Pre-behavioralism in Political Science", in: *American Political Science Review*: 9–31.

Chapter 9
Representative Democracy and the Populist Temptation

9.1 Introduction[1]

My topic sounds rather conventional in terms of a traditional institutional approach. But the 'enlightened neo-institutionalists' of our days came back to the old controversies of the late 1940s when in the United States a debate was waged whether one should introduce a parliamentary system in the USA (cf. von Beyme 1987: 38f). After 1945 even the American Political Science Association—normally refraining from ex-cathedra-normative statements—made contributions about a 'Toward a more responsible Two-Party System' (1950) in order to push the presidential system into another form of representative government, as an American functional equivalent of a British cabinet government. The innovation of neo-institutionalism is aiming at new forms of representative democracies in the light of normative debates on defective democracies in transitional regimes and a more inclusive deliberative or reflexive democracy in consolidated systems—exceeding the typologies of conventional 'liberal democracies'. Liberalism was in many countries undoubtedly the political movement that developed theory and practice of parliamentary government. But though liberalism is still alive, the epithet 'liberal' among radical democrats became a kind of invective.

In the second wave of democratization after World War II the existing forms of representative democracy were so superior that hardly any critic thought that it was necessary to classify also 'defective democracies'. Only one established democratic regime was transformed, the Fourth French Republic in 1958. All the other new democracies showed considerable stability. The very term 'defective democracy' was coined only in the third wave of democratisation. Contemporary typologies of representative democracy include the deficient forms. We are now prepared

[1] This text was fist published by Klaus von Beyme as: "Representative democracy and the populist temptation", in: Alonso, Sonia et al. (eds.): *The Future of Representative Democracy*. Cambridge, Cambridge University Press, 2011: 50–73. The permission to republish this text was granted on 30 April 2013 by Ms. Claire Taylor for Cambridge University Press.

K. von Beyme, *Klaus von Beyme*, SpringerBriefs on Pioneers in Science and Practice 14, DOI: 10.1007/978-3-319-01535-4_9, © The Author(s) 2014

to recognize that also the good old consolidated representative democracies in their genesis were defective. There were early representative forms of government, but they could hardly be called democracies. Even the mother of parliamentary government, Britain, was a latecomer in democratization, if we only look at the time when universal suffrage was achieved—and not only male suffrage as it existed in France still in the twentyth century and in Switzerland until 1971. Even Britain needed a long time from the Samuel Sandy's first move for a more responsible government in 1741 to the year of 1835 when it was finally recognized that no prime minister could govern without the confidence of the parliamentary majority.

Representative regimes existed also in the form of assemblies of estates—as in Sweden until 1866. When this late feudal system was overcome Sweden needed still half a century to implement parliamentary democracy in 1917 (von Beyme 2000a, b: 16ff). The early representative regimes in terms of Braudel's history of 'longue durée' were dubbed "some kind of parliamentarism" (Palme 1969; Turkka 2007: 14ff). But 'parliamentarism' did not yet include full responsible parliamentary government—even if parliaments or estates as in Sweden in the 'Frihetstiden' of the late eighteenth century were preponderant in the system. These regimes were not only deficient democracies, but also defective parliamentary systems. But undoubtedly they were forerunners of the contemporary forms of representative democracy which are the focus of this paper.

There are three theoretical and ideological conflicts combined with representative forms of government:

- Those who fight for a consolidated representative democracy analyze in which elements of their regimes fall short of the normal way of operation in the system and they develop theories of 'defective democracy'. They no longer classify regimes only by formal criteria (parliamentary or presidential representative democracy) but dig into the integration of several subsystems in a system of consistent democratic rules of the game in an 'embedded democracy' (Merkel et al. 2003: 48). Because the early representative systems from the viewpoint of democratic theory were already defective, we have to throw a short view on the forerunners of contemporary forms of representative democracies.
- In *normative theories* the deficiencies of representative democracies are analyzed and visions are developed for 'better democracies' in the spirit of 'Republicanism', deliberative democracy, reflexive democracy and other concepts for the 'democratization of democracy'. An analysis of present forms which does not lose itself in mere institutional enumerations will have to include forms of representative democracy which do not yet exist and serve as a permanent normative challenge for 'transformation' or 'reform'.
- In some countries which developed already a consolidated representative democracy parts of the electorate and small parts of the political entrepreneurs are increasingly unsatisfied with formalist routines of the system. They develop *new populist visions* of better and more representative direct democracy against a mere procedural democracy of elite competition.

9.2 Historical and Contemporary Forms of 'Defective' Representative Democracies

9.2.1 Historical Forms

Representative Democracy is a rather vague catch-all term. According to historical developments the forms of the representative system—not always identical with democracy—have to be differentiated. Representative government in the nineteenth century was the most comprehensive notion in the typologies of constitutions. Even English writers noted that in the German language various words such as *darstellen, vertreten, repräsentieren* coexisted and had different notions about representation, with not only political meanings (Birch 1993: 71). In some cases—as in the work of Benjamin Constant—'representative constitution' and 'constitutional' were almost synonyms.

9.2.2 Terminological Chaos Revisited: Typologies of Representative Democracies

9.2.2.1 De Facto Descriptions

Eighteenth century: Representative government
- Gothic government (Montesquieu)
- Constitute government (Burke)
- Republic (Kant, representative contrary to Rousseau)

Nineteenth century: Landständische Verfassung (constitution of estates after 1815)
- Constitutional monarchy (dualistic)
- Representative democracy (Mill in a negative way)
    - Parliamentary government or responsible government
    - Presidential government (USA)
    - Directorial government (Switzerland)

- Prime-ministerial government (Bagehot)
- Cabinet government (Jennings)
- Party democracy. Parteienstaatliche Massendemokratie (Leibholz)
- Gouvernement d'assemblée (Versammlungsregierung) (Bastid)

Twentieth century Räte-Republik, Council System, Soviets
- Semi-presidential system (Duverger)
- Postdemocracy (Crouch)
- Charismatic media democracy (Korte)

After 1990 Anocracy (Gurr)
(transitional Hybrids (or no hybrids: Przeworski)
Regimes) Defective democracy (Merkel)
Exclusive democracy
Illiberal democracy
Delegative democracy
Democracy of enclaves

Minimal democracy
Façade democracy

9.2.2.2 Normative Types of Desired Representative Systems

Free Soviet system (Anarcho-Syndicalists)
Plebiscitarian democracy (Pateman, Barber)
Republicanism (Pocock)
Deliberative democracy (Habermas)
Dialogic democracy (Giddens)
Reflexive democracy (Schmalz-Bruns)
Subpoliticized democracy (U. Beck)
Cosmopolitan or transnational democracy (Held)

This survey contains only types of representative democracy or representative government which have been widely discussed in the literature. There are, however, also attempts to create more precise differentiations in excessive typologies, such as 'parliamentary investigation state' or 'parliamentary control-state' etc. (Küchenhoff 1967: 881). Constitutionalism and monarchy' were frequently used synonymously. Most interpreters, however, used constitutional government as a subgroup of representative government. It was opposed to absolutist, despotic or bureaucratic form of government—those which Kant called an 'Unform', a 'non-type' of government. Contrary to Rousseau Kant did not identify the Republic with direct democracy: 'Every true Republic is—and cannot be otherwise—a representative system of the people'.

Since 1789 political theory was increasingly divided according to ideological camps. Conservative writers were inclined to identify 'representative government' with 'popular sovereignty' in a negative way. Liberals on the other hand used the term 'representative government' just for the opposite, for systems with elections, but by no means universal suffrage which belongs to the minimal criteria of what might be called 'democratic popular sovereignty. Representative democracy was discussed by Mill and John Stuart (1861), Ch. VII: 256 but he saw two kinds of dangers combined with it: a low grade of intelligence in the representative body, controlled by popular opinion, and class legislation by the numerical majority. The common mode of avoiding these dangers was 'limiting the democratic character of the representation, through a more or less restricted suffrage'. This was feasible for a certain transitional period. Mill's way out of the predicaments of representative democracy was 'proportional electoral law' to create a broad representation of groups and interests.

Representative government among liberals was a popular notion because it excluded an imperative mandate where constituencies could impose concrete orders on how their representative should vote in parliament. Edmund Burke in his famous speech to the electors of Bristol in 1774 (1864: 447) alienated his constituency when he promised 'to live in the strictest union' with his constituants, but mandates 'are things utterly unknown to the laws of this land'. He was not re-elected. John Stuart Mill (1873: 240) even refused to canvass and to answer questions concerning religion. Moreover, he advocated electoral rights for women—which was not very popular at that time. A 'well known literary man', not identified by Mill: 239 with his name, noticed "that the Almighty himself would have no chance of being elected on such a programme".

In France the term 'gouvernement représentatif' was accepted earlier than in Germany. In Germany most writers continued to speak of 'Landstände' (estates)—and thereby avoided the modern connotation of the term 'representation' which meant representatives elected by the voters of the whole country, but with a limited mandate for legislation. Legislation generally was shared with the Crown. Lamennais (1823: 5), in his early days as a religious conservative, still criticized this 'fashion of a notion': "Ce prétendu gouvernement représentatif ne représente rien". This pamphlet showed the typical conservative identification of representative government with constitutional monarchy, and more specific with the British system. Some pamphlet-writers in the era of restoration after 1815 went even further and called the term 'representative government' a *perfidie criminelle*, because it combined a "passive principle of government" with an "active notion of representation' (Réfutation 1816: 72).

The promoters of representative government, on the other hand, transformed the notion into a historical ideology and derived it from the liberties of old Germanic tribes. Some legal schools did this even in Spain with the Visigoths who founded an early civilization on the Iberic pensinsula, or in Russia with the Varagians, invading from Scandinavia, who founded the first 'state' on Russian soil in Novgorod (von Beyme 1965). The legend of 'gothic government' was, however, challenged by political thinkers such as Sieyès, Guizot, Comte or Mohl who practiced themselves the system as parliamentarians. Sieyès was so much socialized in the spirit of the principle of popular sovereignty during the French revolution that he called the British House of Lords" a 'monument of gothic superstition'. This showed, however, that the notion of 'gothic government' was still familiar, though no longer used in a positive way.

'Constitutional monarchy' in the British sense was compatible with parliamentary government with a preponderance of parliament. In the German-speaking countries, however, constitutional monarchy was frequently used as a barrier against parliamentary government. From 1815 to 1918 the idea of a representative government with a strong position of the monarch was advocated, and conservatives from Sweden to Italy have accepted this new form of a hypothetical balance of power, developed in a kind of German–Austrian 'Sonderweg' (special road) of development.

Parliamentary government as a notion was accepted even in Britain rather late and in the light of the doctrine of 'king in parliament' the term was still considered as an insult. From Burke to John Stuart Mill—thinkers in favour of a parliamentary system with preponderance of the legislative—used terms like representative government, constitute government, responsible government or mixed government. In 1834/35 the last conflict between king and parliament was decided in favour of the representatives and the majority rule of the parliamentary system was accepted without further interruptions by actions of the king. Parliamentary government developed a notion for the short time between the two parliamentary reforms of 1832 and 1867, when parliament was preponderant, but without a clear party system. In the era of Disraeli when alternative parties were strengthened, new terms came up, such as 'party government', 'cabinet government' or—since Bagehot 'prime ministerial government'. Most of the terms had ideological connotations and were not always able to survive in the classifications of scholars.

In the second half of the nineteenth century representative government was increasingly taken as the general notion and sub-classified according to types of relationship between parliament and executive. The usual classification differentiates between the monistic system of 'parliamentary government', opposed to dualistic systems such as 'presidential government' (USA) or the *directorial type* (Switzerland). Against a wide-spread error in many newspapers, *semi-presidential regimes* are subtypes of parliamentary government because the popular elected president is confronted with a prime minister who needs the confidence of the parliamentary majority and the president has the counter-weapon of the dissolution of parliament in his hands (von Beyme 2000a, b: 12ff). Since the victory of universal suffrage—in Britain later than in France and Germany—representative government is mostly substituted by the terms 'democracy' or 'representative democracy'.

Representatives are free agents who do not accept mandates. The mandates are developed not so much by the constituencies which are normally quite heterogeneous, but rather by party machines—imposing party discipline with the exception of some moral issues of Weltanschauung. As Hanna Pitkin (1967: 221) formulated in her seminal book on *The Concept of Representation*: "representative government is in reality just party competition for office". This was in tune with Schumpeter's (1942, Ch. 23) reduction of democracy to a 'democratic method' which organizes competition among political leaders. In Germany 'party democracy' or 'parteienstaatliche Massendemokratie' was emphasized (Leibholz 1967: 94) in which representation is reduced to "meetings of party delegates with a mandate'. The elitist concept of democracy for a while dominated the American debate, until 1968 when it became anathema for the new left.

The bitter ideological conflicts about representative government after 1945 withered away. Parliamentary government, in one variation or the other, became the preferred model of all parties. Very few temporarily advocated the alternative of a council (free Soviet) system (von Beyme 2000a, b: 202ff). But dysfunctional developments secured new ideological quarrels about the adequate form of representative government. In several waves the overwhelming role of party government in contemporary forms of representative democracy were criticized.

9.2.3 Contemporary Forms of Defective Representative Democracies

Left-wing extremism among the Communists used to be a major challenge to what they called 'liberal bourgeois democracy'. Since 1989/90 this challenge withered away. Even in Russia we saw the strange example that a former Communist, Khasbulatov (1993), in the Duma, the Russian parliament, taught his deputies lessons about presidential, parliamentary and semi-presidential systems in a conventional way to find guidelines for a future Russian Constitution.

Four species of defective democracies have been differentiated:

- *exclusive democracy* without sufficient control, and exclusion of certain layers and groups from efficient participation.
- *illiberal democracy* with an underdeveloped legal state (*Rechtsstaat*),
- *delegative democracy* which remains semi-pluralistic and has a tendency toward concentration of political power in the often populist executive—which frequently violates parliamentary and judicial institutions,
- *democracy of enclaves* where the effective power remains outside the representative system, mostly with the military (Merkel et al. 2003: 69, 72f).

In all of them a democratic constitution and elections survived. But participation does not lead to all-encompassing representativeness. These shortcomings hint, however, at the necessary balance of 'liberal representative democracies' between representation via participation of all the citizens and the existence of unelected institutions, such as courts and judicial review. These are representative only in the abstract sense that in many countries the more neutral institutions such as constitutional courts have the highest reputation among the citizens. A general rule states that in many defective democracies participation and a certain representation is existing—though elections are not always fair—but the guarantee of liberal principles of the human and social rights are underdeveloped. These models apparently are subtypes of a root concept which can be called 'liberal democracy'.

Most writers accept *hybrids* between liberal democracies and authoritarian regimes. There is something in between both regimes types, such as 'façade democracy' where democratic and representative institutions disguise authoritarianism. Only some authors do not accept 'grey zones' in between. Hybrids are considered as based on a logically inconsistent taxonomy (Przeworski et al. 2000: 57).

However, those authors who consider democratic elections and vertical accountability as a sufficient condition for democracy, fail to distinguish between Sweden, the UK, France on the one side and Russia, Indonesia or Bolivia on the other side, since all governments of those countries are democratically elected.

The distinction of liberal and defective democracies is based on a static status quo analysis. As regime typologies they cannot integrate a dynamic time dimension. Such a time dimension is explicitly considered by 'consolidology' (Schmitter 1995; Linz/Stepan 1996; Croissant/Merkel 2004).

Consolidating regimes were classified in three types, (Beichelt 2001: 143) such as:

- *formally democratic parliamentary systems* (Estonia, Latvia, Slovenia, Czech Republic, Hungary),
- *transitional regimes as balanced systems* (Bulgaria, Macedonia, Moldova, Rumania) and
- *minimal democratic regimes* (Slovakia, Russia, Ukraine). They have been subdivided in 'parliamentary systems' (Slovakia) and 'regimes dominated by a president' (Russia, Ukraine). The latter typology showed how quickly all our taxonomies can be obliterated. Slovakia after some years entered the formally democratic systems and was the first former Communist country except Slovenia to be integrated into the zone of the 'Euro'. In the case of Russia it became doubtful whether it is still 'minimal democratic' or already 'semi-authoritarian'. This is one of the reasons why among an overall comparative analysis of contemporary forms of representative democracies these consolidating systems mostly play a marginal role—perhaps with the exception of the Czech Republic, Slovenia and Hungary. This is the more so, as representativeness initially was manipulated by 'electoral engineering' as a subtype of 'institutional engineering' (Krohn 2003: 61). These processes created unstable electoral systems and volatile voting behaviour which cast some doubts on the 'representative' character of some of the new democracies.

The great transformation of dictatorships into democracies in the third wave brought back the old debates. Juan Linz (1990) came up with the idea, that presidential systems ruined democracy (especially in Latin America), Dieter Nohlen (Nohlen/Fernández 1991) and others were able to show that the ruin of democracy was not simply due to the form of representative government, but party systems have to be taken into consideration and that also parliamentary and semi-presidential governments declined under certain social and political circumstances (which we knew already from the fate of Italy, Austria, and the Weimar Republic between the World Wars).

Eastern Europe was the great field of experiments with representative government. Again the empirical results have been classified in a normative way: parliamentary systems fared better than semi-presidential regimes (Lijphart 1992, 1994). In fact the causality is rather the other way around: socioeconomically developed societies and politically more mature systems, such as the Czech Republic or Hungary have chosen parliamentary government, whereas less developed countries preferred semi-presidentialism which proved among the mature Western systems efficiently working only in France. Comparison of East-European systems, endorsed rather the Linz-hypothesis that parliamentary systems developed in a more democratic way than some semi-presidential systems (Merkel et al. 1996: 85).

9.3 Neo-normative Models of Representative Democracies

New notions and ideals are spreading in the literature about the decline of traditional representative democracy:

- *Negative connotations* are inherent in the terms of post-democracy.
- *Positive connotations* are transported by the terms 'Republicanism', deliberative democracy (Habermas) or dialogic democracy (Giddens).

In the age of *post-democracy* the elites receive less deference and "the secrets of politicians are laid bare to the democratic gaze". Nevertheless virtually all the formal components of representative democracy survived (Crouch 2005: 12, 22).

Neo-normative thinking discovered that the integration of citizens by a common social morality for a democratic society is more important than the institutional integration: 22. Republicanism having an old tradition since the Renaissance was—less successful than liberalism—recently rediscovered. Republicanism and liberalism are frequently considered as incompatible oppositional elements. One advantage of Republicanism seemed to be that it was more democratic than liberalism—but more liberal than the traditional (representative) democracy (Llanque 2003: 7). Republicanism aimed at the revitalization of citizenship in decentralized self-government and was—since Hannah Arendt—directed against: a state apparatus and legitimation via 'nationalized' and state-funded parties, and an a-political privatism of a depolitized population.

Republicanism sees an inevitable connection between citizen's virtues and self-government. Liberalism does not exclude such a combination, but considers it as accidental (Sandel 1995: 57). How intimate this relationship should be? Habermas (1992: 360f) already doubted the feasibility of a polemical separation of civil society and state since a balancing of interests and powers needs the framework of a rule of law (*Rechtsstaat*). Discourses in the model of deliberative democracy have stronger normative components than the liberal state, but weaker components than the republican ideal. Both concepts start from a vision of a 'whole':

- *liberalism* postulates in a rather abstract way that the Constitution is primordial,
- *republicanism*, on the other, starts from the sovereign citizenship.

The discourse in the model of deliberative democracy starts from a decentralized society which creates a political public sphere in order to deliberate about the various social problems and interests.

Deliberative democracy is a normative hope, but post-democracy did not come close to it. Leftist writers complain that there is hardly any fundamental criticism of capitalism—but only denunciation of abuses by transnational multinational corporations. Many populists, such as Blocher's peoples party in Switzerland, are rather capitalist-minded and offer no hope for fundamental change of the system. New types of democracy are developed every day: 'democracy of negotiation', 'civil society', 'cooperation of net-works', 'subpolitisation' (Ulrich Beck), 'cosmopolitan democracy' or 'transnational democracy', a combination of Habermas and theories of civil society (Held 1993), 'reflexive democracy' against centralized and homogeneous structures of decision-making and in favour for various layers of processes of decision-making which are connected by networks (Schmalz-Bruns 1995: 164).

But even writers who claim to have shown that 'our whole political knowledge is obsolete, especially the concept of representative democracy', started polemics against all the attempts since John Rawls to revitalize a public ethic in the tradition of Kant. Democracy should be re-installed in its 'laicist functions of organization for special interests, conflict resolution and guarantees for civil liberties' (Zolo 1997: 217). Other authors are even more violently opposed to the 'tyranny of common sense', proclaimed by 'communitarian lay-priests' and vote for an 'egoistic society' (Herzinger 1997: 61). One camp hailed the 'comeback of the citizens'—the other one the 'true representation of egoistic interests' in a 'realistic model of democracy'.

If we look at former leftists like Hardt and Negri (2002) in their vision of 'Empire', there is hardly any hope for a fundamental change. Foucault's spirit is spreading. Each power structure bears inherently its counter-power. Populism is offered by some more sophisticated leaders as an element of Foucault's counter-power. The more liberal democracy was universalized after 1989/90, when communism and other dictatorships collapsed, the more liberal democracy—which is essentially a system of representation—is considered as a rather limited system.

Torchbearers of 'radical politics', such as Anthony Giddens (1994: 112), developed the idea of a 'dialogic democracy', hoping that this concept will be more than an extension of liberal democracy. It should create—not new rights and representations of interests as in representative democracy—but further cultural cosmopolitism, decisive for reconstructions of social solidarity.

Social solidarity is said to be 'reconstructed' by many new social movements, but it is no longer encompassing and universal. Identity politics became a catchword of postmodern democracy and changed the connotations of representation. But prospects according to Colin Crouch (2005: 119) are slim: "Nor will populism be contested by trying to move beyond identity politics to a Third Way political appeal which tries to evade the very idea of identity". But political parties which claim to represent the masses of people need to do so by articulating an identity for those people (Pizzorno 1993). The more these identities are artificially 're-constructed', the more other possible identities are neglected. The established parties have been compared to large corporations: both avoid risks, the corporations do not like risky investments, the established parties try to omit investing in identity-building for new social movements (Crouch 2005: 120). Catch-all parties prefer to cooperate with selected social movements, but shy away from the necessary specialisations of populist movements. New social movements mostly were only successful when they accepted cooperation with established interest groups and parties—as the ecological or the feminist movements have shown.

Populist movements—as other new social movements—cause sometimes 'elite-directing forms of politics' against former "elite-directed politics" (Inglehart 1990:

338). But this does not mean that populist movements have no permanent impact. Since the 'participatory revolution' in the 1970s and 1980s electoral competition has changed: medium range, non-encompassing ideologies, special issues and a growing impact of individual candidates and their profiles prevailed (Kaase 1984).

New terms gave more respectability also to some populist movements: they claimed to represent the 'civil society' against the 'political class'. But no movement has ever permanently incorporated the civil society. Some critics (Latour 1991: 68, 188) already believe that modern constitutions have become a victim of their success and are about to collapse. Mobilisation of collective groups has created so many hybrid forms that the constitutional framework can no longer keep them together. In the light of postmodern 'normalisations' this is certainly an exaggeration. The production of hybrids which explicitly and collectively will be part of a 'non-modern constitution' and an 'enlarged democracy' is a utopia on the basis of 'reunification of nature and society' which is not under way and so far remains a hope of the ecological movement.

Even a normative thinker, such as Habermas (1992: 446), fighting for deliberative democracy admitted that civil society is always in danger to degenerate:

- by populist movements which defend traditions and identities against a capitalist modernization,
- by movements which exceed influence-seeking and try to transform into power organizations,
- and by social revolutions which re-establish a historical subject in a teleological theory as torchbearer of progress.

But even this realistic view on deliberative democracy found severe critics, such as Richard Rorty (1989). Habermas in his view remained a 'metaphysical thinker' because he continues to hope for 'consensus via discourses'. But he shared with Habermas his commitment for the 'citoyen engagé'.

9.4 Populism as Challenge to Representative Democracy

Liberal representative forms of government were not only challenged in new democracies and by normative political theories. More dangerous were the challenges by new social movements—mostly populist. Many consolidated parliamentary party governments in Western Europe have been increasingly challenged by right wing populist attacks, ranging from Haider's movement Austria, Le Pen's National Front in France to Berlusconi's Italy. Even in Scandinavia right wing populist parties emerged. They attacked not only tax, welfare or immigration policies, but also the mode of governing by established party cartels. In the 'Third World' left wing populism under Chavez, Morales and others is protesting against social exclusion and ethnic marginalization of the indigenous population. In Europe right-wing populism was prevailing, protesting against too much inclusion of immigrants and politically active minorities. With respect to representative democracy it should be noted, however, that these right populists movements normally do not challenge the basic democratic and representative institutions of the system.

Populism started as a kind mystical union of the people—rather a syndrome than an ideology (Wiles in: Ionescu/Gellner 1969: 166). Leaders boasted of a direct communication with the people. Populists think in terms of 'social movements'—with some exceptions such as *Forza Italia*—rather than organizing a party. The basic creeds are:

Populist propaganda is *less programmatic and more moralistic*. Since populists quite frequently pretend to o science and its inhuman rationalism they appeal to common prejudices in the people and they hardly ever participate in critical controversial debates. They prefer myths of conspiracy: 'we have been cheated' or 'we have been neglected by the establishment' are popular slogans. 'Virtue' resides in the simple people and their collective traditions.

Populists *fight the* 'corruption' *of the established elites* which are declared as no longer representative of the people. They prefer the term 'political class' instead of the positive connotations of the notion 'elite'. In third world countries primitivism (such as the mystification of Aztec heritage in Mexico) merges with progressivism, close to socialist ideas. There is rarely a consistent doctrine—sometimes the populists started as a single-issue-movement, which creates not a system of related creeds as in ideology, but only a stubborn overestimation of one issue in society.

The ideas of populists on representative government depend on their location in the left–right-scheme. There is still a debate whether all populist movements are right-wing, just the more moderate form of right-wing extremism. Recently left-wing populists have been discovered (or re-discovered because they existed in Russia and in the United States already in the nineteenth century): Originally the Green movements were considered as populists and sometimes the post-communist parties, such as the *PDS/Die Linke* in Germany which became an important party for representing East German interests.

The notion of populism should be differentiated from extremist movements—though they, such as the fascists, have many populist features. There are, however, neo-fascist parties—such as the NPD in Germany—which show little populist features and behave like fascists in a rather bourgeois outfit. Most extremists dislike increasingly to be dubbed as 'fascist'. But populism is taken as a kind of honorary title, as Haider (1994: 57) wrote in his book *Freiheit, die ich meine* (The liberty I have in mind) that populism is a necessary movement in democracy in the fight 'against the commands from the ivory tower of the political class' and its 'disgust of the people'.

Seymour Martin Lipset in his Political Man (1960) was the first to discover 'extremism of the centre'. But he had mostly third world countries in mind. In Latin America left-wing populists were wide-spread. With growing Europeanization and globalization populism in the liberal-conservative centre of the party spectrum became more frequent—from Poujade in France to Glistrup in Denmark or Blocher in Switzerland. The losers of recent economic developments turn to be populists. They blame within the regime a 'degenerated representative system' with its elitist cartels and denounce cooperation with certain scapegoats from the European bureaucracy in Brussels to foreign investors 'invading' the country and their neo-liberal ideology, the CIA or even the United States as a whole. Populists normally fight against three enemies:

- big industry,
- trade unions—especially when they cooperate with big industry in corporatist institutions,
- the 'state' of the established elites which tries to bring big interests to cooperation. The state is no longer the target of demands for alimentation, but populists rather claim to bring the state back to an orientation towards the 'common good'—which normally is defined rather vaguely.

Instead of the traditional party state populists favour a society of free associations—in the French leftist tradition of Proudhon or British guild socialism, in the more conservative tradition closer to ideas of Otto von Gierke (Priester 2007: 220).

The opposition to 'big corporations' does not mean that all populists fight against capitalism, and not even the neo-liberal version of it. Many populists could be called 'anarcho-liberal'. On the contrary: from Glistrup in Denmark to Haider's (1994: 150ff) FPÖ in Austria and Blocher's SVP in Switzerland one kind of conservative populism is predominantly directed against the welfare state. Many populists have accepted the logic of individualization. This is another reason why they prefer rather loose networks instead of organized parties. Political 'entrepreneurs' prevail among the leaders of populist movements, such as Poujade, Fortuyn, Haider. Berlusconi is certainly the prototype of this type of a political 'parvenu' who benefited from a degenerating party system in Italy. Only a minority of populists today can be identified with right-wing extremists. Initially this was done in the case of Le Pen in France. But even this most enduring leader of a populist movement as an 'elder statesman' tried to agitate in a more respectable and centrist way.

Extremist movements tend to belief in the overthrow (or at least radical change) of the existing system and do not recognize the constitutional rules, whereas most populist movements grudgingly accept the rules and want to change only minor elements: mostly the electoral law and they demand the direct election of the president. When this postulate is already implemented in the system, as in Austria, Haider (1994: 235) complained about a costly duplication of offices, without claiming a fully presidential system. Populist movements are rarely revolutionary. They continue to hope to bring the establishment back on the right road to democracy and proclaim a 'Second Republic' in Italy, a 'Third Republic' in Austria (Haider 1994: 189) and a 'Fourth Republic' in Poland under Kaczynski.

Originally populism was a rural movement. In the era of globalization it tends to turn into an urban phenomenon. Competition of foreign migrants was always a breeding ground for populists in cities of the United States. The new populists claim that the native people is alienated and doomed to decline. Populists pretend to represent the national population. Only leftist populists sometimes claim to be fundamentally opposed to right-wing populist ideologies. But already Proudhon as foe of Napoleon III proved at one point of his career after 1848 to be ready to cooperate with Bonaparte and was deeply disappointed when the president of the 2nd French Republic and from 1851 on the Emperor, did not accept him as a political partner. One issue which today separates right and leftwing populists is immigration. Ecological populists favoured more immigration and a multi-cultural society which was abhorred by right-wing populists (Betz 1994: 179ff). Both shades of populism are no nationalists in the old sense and favour regional autonomy and decentralization.

Populism is organized by charismatic leaders such as Poujade or Le Pen in France. The leader claims a special title to 'represent' the needs of the people. If this charisma fails or is substituted by bureaucratic leadership and—what Max Weber called 'Veralltäglichung des Charismas', 'routinization' and decline of the special attraction of the leader—the populist movement disintegrates very quickly. In Germany the fall-down of Schönhuber, the founder and leader of the populist 'Republicans' made them quickly marginal in the system. Similar signs of disintegration recently were experienced by Blocher in Switzerland, when his movements did no longer follow him in his policy towards the government. When Pim Fortuyn was killed in Holland he was not easily substituted in his movement. In many countries the 'intellectualisation' of leadership was not successful and erosion was the consequence when the masses get bored by the ever repeating slogans (cf. Stöss 2000: 178). A lack of professionalization in parliaments proved to be detrimental to the growth of populist movements in the long run. When populist policy styles are adopted even by the big parties the populist smaller groups do no longer have an advantage.

The routinization of populist movements starts when they get close to power. Many of them prefer to remain in opposition to 'keep clean' the purity of their basic creeds. Nothing is more compromising than being held responsible for bad policies, such as Haider in the Austrian government, Gregor Gysy as left-wing populist in the Berlin government, or the support of populists for bourgeois governments in the Benelux or the Scandinavian countries.

A unique case in Western Europe was Berlusconi who reshuffled the whole Italian party system in the early 1990s. When his 'Second Republic of Italy' proved to be even more corrupt and undemocratic than the former 'classe politica' Berlusconi was toppled. He was able to survive for a while only with the help of a democratized neo-fascist movement and a regional populist group such as *Lega Lombarda*. The case is unique in European history because Berlusconi—accused for corruption and determined to change the law in his favour—managed to be reelected in 2008.

Coalitions are always shaky, coalitions of populists, however, are even more likely to be unstable. Some great electoral successes proved to end up in disasters because the populists lacked professional cadres in order to act successfully in parliaments, as the NPD (National-Demokratische Partei Deutschlands), the DVU (Deutsche Volksunion) or the 'Republicans' which were temporarily represented, showed in German Laender diets (Holtmann 2002).

The enlightened followers of old modernisation theories—such as Przeworski (2000: 187ff)—were inclined to correlate the stability of representative democracy with levels of economic or demographic growth. But populism showed that there is a subjective side of the coin: groups which are bored and dissatisfied with democracy and are complaining on a high level of economic performance in the system, since there are always losers in rapid processes of change.

Since the 1980s a normative debate about the virtues and failures of populism developed. The established parties tended to consider populist movements as bad. In Germany—a country which invented the constitutional possibility to outlaw a party, in the meantime copied by other countries, such as Russia—even prohibition was considered for new unwelcome competitors. But in the age of the new social movements the creative forces of populists have been discovered—even by formerly leftist authors (Priester 2007: 220). Populist leaders are classified to the extent they developed representative democracy in a seminal way: De Gaulle and Gandhi were the 'good populists'—militant regional populists, such as Sinn Fain or the Bask leaders, were the 'bad populists' —as long as they supported terrorism.

With the spread of terrorism all over the world the basic criterion for classification of new movements is whether they accept or refuse terrorism as a tool of political conflicts. Terrorist action is the very opposite of all kinds of representative democracy since it is neither deliberative, representative or democratic. Additional criteria for the differentiation of right-wing extremists from radical right-wing populists are a consistent and continuous anti-Americanism and Anti-Semitism—not part of most populist ideologies.

The initial verdict against the populist revolt against representative democracy has been mitigated by several experiences:

(1) *Populist styles captured the leadership* in established parties and changed the rules of representative government—as has been demonstrated for leaders such as Blair in Britain or Schroeder in Germany. Charismatic media democracy created a populist style in conventional politics (Korte 2003).
(2) Populism in most West-European systems was *no threat to representative democracy* so far. In most West European systems the populists did not exceed 10 % of the votes, with the exception of the Front National in France, the FPÖ in Austria, Blocher's SVP in Switzerland and the Norwegian Fremskrittspartiet. The SVP became the strongest party with 27 % and Blocher was able to overthrow the magic formula in the distribution of seven governmental seats. The SVP got a second seat, the 'proportional system', however, was maintained. Populists sometimes have enormous success, but they are never safe. The fluctuations are enormous (data in: Betz 1994: 3), as the Poujade movement showed in the Fourth French Republic which withered away in a few years.

In the 1980s the social movements were suspected to substitute the old institutions. However, the final result was an integration into the system. The new social movements were mostly successful in agenda-setting and provocation of new issues in the public debate but no even in launching amendments to the

Constitution. Most populist movements were flash parties. Only leaders who turned to authoritarian rule (Peron in Argentina), or successfully tried to occupy the centre because the alternatives had withered away and substituted their populist image by the attitude of 'statesmanship' (Berlusconi in Italy) were able to remain a decisive factor in the representative system of their country.

(3) Two variations of a kind 'inbuilt populism' in contemporary representative democracies have been differentiated (Decker 2006a, b: 22, 26):

- *Moderate populists* accept the constitutional-representative model of democracy, and strengthen it by emphasizing more inclusion of groups and interest and a deliberative democracy.
- *More radical populists* favour plebiscitarian democracy. 'Decisionism', on the basis of a unitary will of the people, substitutes deliberation.

The second version of plebiscitarian democracy model might be a potential danger, but the drive of the parties challenging the system—with the exception of Italy—was never strong enough to change the system and its institutions. Germany proved to be particularly protected against right-wing extremism and populism because of

- its Nazi past,
- because the two major parties are moderately welfare oriented, and merged their forces twice in a Grand Coalition to reform the system.

Populist slogans in catch-all parties are increasingly stolen from the populist groups. Populists finally remained 'a-political' because they don't like compromise as long as they remain in opposition. Populists pretend to mobilize. But frequently the result was manipulated 'pseudo-participation'. As soon as populists are established, they learn to work in terms of compromises with other groups. When they are accepted parties they loose their uniqueness. It happened to some progressive parties on the right and to the green parties on the left of the spectrum of parties.

9.5 Conclusion: 'Direct Plebiscitarian Democracy'—The Way Out of Liberal Representative, Merely Procedural Democracy?

Representative democracy—by liberals and their enemies dubbed 'liberal democracy'—is torn to pieces between utopian demands and functional reality. The rhetoric of democracy got weak since 1990 when the system remained unchallenged without authoritarian alternatives. Representative democracy by many populists such as Haider or Blocher is said to be 'minimalistic' and does not invite any enthusiasm among the citizens. Even defenders of liberal democracy like Andrew Levine (1981: 7), have admitted: "Indeed, I will go on to suggest that the political institutions proper to liberal democratic politics, representative government and the party system, so far from implementing democratic values may actually betray them." In this perception liberal democrats pretend to be liberals and democrats. The theory to which they are committed is genuinely liberal but not—at least in its practical implementation—genuinely democratic. Representation included mostly the notion of independent legislators who follow their consciousness and instincts rather the changing popular moods in a *Stimmungsdemokratie* (democracy of the popular moods).

Direct democracy and popular legislation are frequently demanded, and even accepted by the radical wing of liberal democrats from Carol Pateman (1970: 111) to Benjamin Barber (1984: 10, 33).

There is a stubborn believe that direct democracy is not a question whether it should be organized, but only how it can be implemented (Decker 2006a: 5). Non-normative scholars, however, try to answer the question empirically whether direct democracy and legislation via referenda with a deliberative discourse is really better and more efficient. About 550 popular initiatives and referenda have been examined. The result was not encouraging: no Habermasian 'power-free discourse' has been discovered. The results of referenda were mostly middle class- and status quo-oriented. Radical groups or even semi-leftist groups, such as trade-unions, normally failed to gain anything from popular legislation. The growth of the welfare state has been promoted rather by representative parliamentarians in Northern Europe than by direct democracy in Switzerland (Wagschal in: Freitag/Wagschal 2007: 326f; Moser/Obinger ibidem.: 357). More innovative legislation is not expected even by some other experts (Kranenpohl 2006: 38). The 'independence of representatives' remained the result of the impracticability of direct democracy. Representatives in many cases had to decide after a failure of popular votes (switching from left-hand traffic to right-hand traffic in Sweden, or lowering voting age to 18 years in Denmark). Oddly enough, Brown and Sarkozy, leaders in two major countries that faced hostile majorities to the EU constitution, recently confessed in October 2007 that there will be no referendum on this question next time. Referenda serve sometimes a complementary institution of advice. In Italy consultative referenda frequently produced rather reasonable results which served as a guideline for the representative decision-makers. Unexpectedly even subsidizing party finances by the state was admitted by a small majority. Sometimes an equal rank of representative and direct democratic legislation is envisaged. This is, however, very difficult under two circumstances:

- In a *federalist system* there is the danger that the federal chamber gets meaningless. Even in Switzerland referenda develop centralizing impacts.
- In a *system with sovereignty of the constitution and judicial review by a constitutional court*. In such a system ex-ante-judgements of the court have been recommended in order not to violate the "sovereignty of the people" (Kranenpohl 2003).

Empirical doubts about the efficiency of direct democracy never convinced normative theoreticians. They emphasize—not without good reasons—that direct democracy has impacts on political socialization and learning which in the long run stabilize also representative democracy (Waschkuhn 1998: 514).

We end up in a *paradox of postmodernism:* in 1968 the criticism of liberal democracy took violent forms in terms of a radical normative theory. Forty years later normative theory—which became more realistic in the meantime and learned from the criticism of the 1968 generation—is confronted with the same objections as representative democracy in 1968: that it represents hardly more than a normatively embellished duplication of a dreadful political reality. Science seems to follow the arts: 'kitchen-sink-art'—and other variations of postmodern art—do not lead back to a higher ideal of normative aesthetics but deliberately lead back to a miserable every-day life which is grudgingly accepted.

References

Barber, Benjamin, 1984: *Strong Democracy* (Berkeley: University of California Press); (German edition, 1994: *Starke Demokratie. Über die Teilhabe am Politischen* (Hamburg: Rotbuch Verlag)).

Beichelt, Tim, 2001: *Demokratische Konsolidierung im postsozialistischen Europa. Die Rolle der politischen Institutionen* (Opladen: Leske & Budrich).

Betz, Hans-Georg, 1994: *Radical Right-Wing Populism in Western Europe* (Houndsmills: Macmillan).

Beyme, Klaus von, 1965: "Repräsentatives und parlamentarisches Regierungssystem. Eine begriffsgeschichtliche Analyse", in: *Politische Vierteljahresschrift*: 145–159.

Beyme, Klaus von, 1987: *America as a Model: The Impact of Democracy in the World* (Aldershot: Gower).

Beyme, Klaus von, 1988: "Right-Wing Extremism in Post-War Europe", in: *Westeuropean Politics*: 1–18.

Beyme, Klaus von, 2000, ^{2}2002: *Parteien im Wandel. Von den Volksparteien zu den professionalisierten Wählerparteien* (Opladen: Westdeutscher Verlag).

Beyme, Klaus von, 2000a: *Parlimentary Democracy: Democratization, Destabilization, Reconsolidation, 1789–1999* (Houndsmill: Macmillan).

Beyme, Klaus von, 2013: *Von der Postdemokratie zur Neodemokratie* (Wiesbaden: Springer VS).

Birch, Anthony H., 1993: *The Concepts and Theories of Modern Democracy* (London: Routledge).

Burke, Edmund, 1864: "To the Electors of Bristol (1774)", in: *Burke: Works*, Vol. 1 (London: Henry G. Bohn): 442–449.

Croissant, Aurel/Merkel, Wolfgang (Eds.), 2004: "Consolidated or Defective Democracy? Problems of Regime Change", in: *Democratization*, 11, 5.

Crouch, Colin, 2005: *Post-Democracy* (Cambridge: Polity Press).

Decker, Frank (Ed.), 2006: *Populismus. Gefahr für die Demokratie oder nützliches Korrektiv?* (Wiesbaden: Verlag für Sozialwissenschaften).

Decker, Frank, 2006a: "Direkte Demokratie im deutschen 'Parteienbundesstaat'", in: *APuZ*, 10: 3–10.

Faltin, I., 1990: *Norm, Milieu, politische Kultur* (Wiesbaden: DUV).

Forst, Rainer, 2007: *Das Recht der Rechtfertigung* (Frankfurt: Suhrkamp).

Freitag, Markus; Wagschal, Uwe (Eds.), 2007: *Direkte Demokratie. Bestandsaufnahmen und Wirkungen im internationalen Vergleich* (Münster: LIT).

Leibholz, Gerhard, ^{3}1967: *Strukturprobleme der modernen Demokratie* (Karlsruhe: C. F. Müller).

Giddens, Anthony, 1994: *Beyond Left and Right. The Future of Radical Politics* (Cambridge: Polity).

Gurr, Ted et al., 1991: "The Transformation of the Western State: The Growth of Democracy, Autocracy and State Power Since 1800", in: Inkeles, Alex (Ed.): *On Measuring Democracy* (New Brunswick: Transaction): 69–104.

Habermas, Jürgen, 1992: *Faktizität und Geltung* (Frankfurt: Suhrkamp).

Haider, Jörg, 1994: *Die Freiheit, die ich meine* (Frankfurt: Ullstein).

Hartleb, Florian, 2004: *Rechts- und Linkspopulismus* (Wiesbaden: Verlag für Sozialwissenschaften).

Held, David, 1993: *Political Theory Today* (Oxford: Polity).

Herzinger, Richard, 1997: *Die Tyrannei des Gemeinsinns. Ein Bekenntnis zur egoistischen Gesellschaft* (Berlin: Rowohlt).

Holtmann, Everhard, 2002: *Die angepassten Provokateure. Aufstieg und Niedergang der rechtsextremen DVU als Protestpartei im polarisierten Parteiensystem Sachsen-Anhalts* (Opladen: Westdeutscher Verlag).

Inglehart, Ronald, 1990: *Culture Shift in Advance Industrial Society* (Princeton: Princeton University Press).

Ionescu, Ghita; Gellner, Ernest (Eds.), 1969: *Populism. Its Meanings and National Characteristics* (London: Weidenfeld & Nicolson).

Kaase, Max, 1984: "The Challenge of the 'Participatory Revolution' in Pluralist Democracies", in: *International Political Science Review*, 5,3: 299–318.

Kant, Immanuel, 1957–1964: in: Weisschädel (Ed.): *Immanuel Kant Werke*, Vol. 6 (Darmstadt: Wissenschaftliche Buchgesellschaft).

Khasbulatov, R.I., 1993: "Kakaya vlast' nuzhna Rossii", in: *Sotsiologicheskie issledovaniya*, 11: 18–31.

Korte, Karl-Rudolf, 2003: "Populismus als Regierungsstil", in: Werz, Nikolaus (Ed.): *Populismus* (Opladen: Leske & Budrich): 209–222.

Kranenpohl, Uwe, 2003: "Verkürzen Verfassungsrichter Volksrechte?", in: *Gesellschaft – Wirtschaft – Politik*, 52: 37–46.

Kranenpohl, Uwe, 2006: "Bewältigung des Reformstaus durch direkte Demokratie?", in: *APuZ* 10: 32–38.

Krohn, Tobias, 2003: *Die Genese von Wahlsystemen in Transitionsprozessen. Portugal, Spanien, Polen und Tschechien im Vergleich* (Opladen: Leske & Budrich).

Küchenhoff, Erich, 1967: *Möglichkeiten und Grenzen begrifflicher Klarheit in der Staatsformenlehre* (Berlin: Duncker & Humblot).

Lamennais, 1823: *Extraits du drapeau blanc. Plan d'un libre intitulé : Observations sur le gouvernment de l'Angleterre, dit gouvernement représentatif constitutionnel* (Paris).

Laski, Harold, 1944: "The Parliamentary and Presidential Systems", in: *Public Administration Review*, Autumn: 347–359.

Latour, Bruno, 1991: *Nous n'avons jamais été modernes* (Paris: La Découverte) (German edition, 1995: *Wir sind nie modern gewesen. Versuch einer symmetrischen Anthropologie* (Berlin: Akademie Verlag).

Levine, Andrew, 1981: *Liberal Democracy. A Critique of Its Theory* (New York: Columbia University Press).

Lijphart, Arend (Ed.), 1992: *Parliamentary versus Presidential Government* (Oxford: Oxford University Press).

Lijphart, Arend, 1994: *Electoral Systems and Party Systems: A Study of Twenty-Seven Democracies* (Oxford: Oxford University Press).

Linz, Juan, 1990: "The Perils of Presidentialism", in: *Journal of Democracy*, 4: 51–69.

Linz, Juan; Stepan, Alfred, 1996: *Problems of Democratic Transition and Consolidation. Southern Europe, South America, and Post-Communist Europe* (Baltimore: Johns Hopkins University Press).

Lipset, Seymour Martin, 1960: *Political Man* (London: Mercury Books).

Llanque-Kurps, M., 2003: *Klassischer Republikanismus und moderner Verfassungsstaat* (HU Berlin: Habilitationsschrift).

Hardt, M.; Negri, A., 2002: *Empire – die neue Weltordnung* (Frankfurt: Campus).

Mair, Peter, 2002: "Populist Democracy vs Party Democracy", in: Mény, Yves; Surel, Yves (Eds.), 2002: *Democracies and the Populist Challenge* (Houndsmill: Macmillan): 139–154.
Mény, Yves; Surel, Yves (Eds.), 2002: *Democracies and the Populist Challenge* (Houndsmill: Macmillan).
Merkel, Wolfgang et al. (Eds.), 1996: *Systemwechsel 2. Die Institutionalisierung der Demokratie* (Opladen: Leske & Budrich).
Merkel, Wolfgang et al., 2003: *Defekte Demokratie*, Vol. 1: Theorie (Opladen: Leske & Budrich).
Mill, John Stuart, 1861 (1910, 1960): *Representative Government* (London: Dent).
Mill, John Stuart, 1873 (1928, 1958): *Autobiography* (London: Oxford University Press).
Mulgan, G., 1994: *Politics in an Antipolitical Age* (Cambridge: Polity).
Nohlen, Dieter; Fernández, Mario (Eds.), 1991: *Presidencialismo versus Parlamentarismo* (Cáracas: Nueva Sociedad).
Palme, Sven Ulrich, 1969: "Vom Absolutismus zum Parlamentarismus in Schweden", in: Gerhard, Dieterich (Ed.): *Stände-Vertretungen in Europa im 17. und 18. Jahrhundert* (Göttingen: Vandenhoeck & Ruprecht): 368–397.
Réfutation de la doctrine de Montesquieu sur la balance des pouvoir (Paris) 1816.
Pateman, Carol, 1970: *Participation and Democratic Theory* (Cambridge: Cambridge University Press).
Pitkin, Hanna F., 1967: *The Concept of Representation* (Berkeley: University of California Press).
Pizzorno, Alessandro, 1993: *Le radici della politica assoluta e altri saggi* (Milan: Feltrinelli).
Price, Don K., 1944: "A Response to Mr. Lask", in: *Public Administration Review*, Autumn: 360–363.
Priester, Karin, 2007: *Populismus: Historische und aktuelle Erscheinungsformen* (Frankfurt: Campus).
Przeworski, Adam et al., 2000: *Democracy and Development. Political Institutions and Well-Being in the World, 1950–1990* (Cambridge: Cambridge University Press).
Stöss, Richard, ^{3}2000: *Rechtsextremismus im vereinten Deutschland* (Berlin: Friedrich Ebert Stiftung).
Rorty, Richard, 1989: *Contingency, Irony, and Solidarity* (Cambridge: Cambridge University Press).
Sandel, Michael J., 1995: *Liberalismus und Republikanismus: von der Notwendigkeit der Bürgertugend* (Vienna: Passagen; Wien: Passagen-Verlag).
Schmalz-Bruns, Rainer, 1995: *Reflexive Demokratie. Die demokratische Transformation moderner Politik* (Baden-Baden: Nomos).
Schmitter, Philippe, 1995: "The Consolidation of Political Democracies", in: Pridham, Geoffrey (Ed.): *Transitions to Democracy: Comparative Perspectives from Southern Europe, Latin America and Eastern Europe* (Dartmouth: Aldershot): 535–569.
Schumpeter, Joseph A., 1942: *Capitalism, Socialism and Democracy* (New York: Harper).
Taggart, P., 1995: "New Populist Parties in Western Europe", in: *West European Politics*, *Toward a more Responsible Two-Party-System*. A Report of the Committee on Political Parties of the APSA. APSR, 1950, Supplement 88.
Turkka, Tapani, 2007: *The Origins of Parliamentarism: A Study of Sandy's Motion* (Baden-Baden: Nomos).
Vorländer, Hans, ^{2}2010: *Demokratie* (München: Beck).
Walzer, Michael, 1992: *Zivile Gesellschaft und amerikanische Demokratie* (Berlin: Rotbuch Verlag).
Waschkuhn, Arno, 1998: *Demokratietheorien* (München: Oldenbourg).
Werz, Nikolaus (Ed.), 2003: *Populismus* (Opladen: Leske & Budrich).
Wiles, Peter, 1969: "A Syndrome, Not a Doctrine. Some Elementary Theses on Populism", in: Ionescu, Ghita; Gellner, Ernest (Eds.): *Populism: Its Meanings and National Characteristics* (London: Weidenfeld & Nicolson): 166–179.
Zolo, Danilo, 1997: *Die demokratische Fürstenherrschaft. Für eine realistische Theorie der Politik* (Göttingen: Steil).

Chapter 10
Political Scientists as Advisors to Politicians

10.1 Scholars as Political Advisors: My Personal Experience[1]

In the entrance hall of the Humboldt University in Berlin you can still find a huge inscription on red marble taken from Karl Marx's Theses on Feuerbach, number 11: "Philosophers have interpreted the world in different ways—it is necessary, however, to change the world". This meant that social scientists and philosophers should certainly play a certain role in changing the world. It was a wise decision to defend the inscription against those Western committees who wanted to purge all remnants of Marxism in the former GDR. Nevertheless, in the light of postmodern society, this famous sentence is wrong. It should rather read: "Philosophers continue to interpret the world in different ways, but the world changes rather in a unified direction—though scientific theories have little influence on this process".

In my study of political theories in Europe in the era of ideologies from 1789 to 1914, I counted among 170 eminent political thinkers from Madrid to Moscow ninety-seven who attempted to work as politicians (von Beyme 2002: 949). This was not restricted to leftist thinkers, from Godwin to Proudhon, who tried to work in parliaments. Conservative and liberal thinkers such as Burke, Chateaubriand, Constant, Tocqueville and even Max Weber (as an unelected candidate) wanted to serve in parliament. Today the separation of the scientific and political subsystems makes this kind of representation rather rare.

But during the time of transition to democracy in Eastern Europe after 1989, for a while theories of civil society urged many scholars and artists to spring into the political arena—from Vaclav Havel in Prague to Andrei Plešu in Bucarest. Most of these new elites were quickly frustrated, the apparatchiki came back.

Rarely was a political scientist with a good reputation dating from communist times, like our Polish colleague Jerzy Wiatr in Poland, able to play a role in the

[1] This text was first presented as "Political Advisors to Politicians (Presentation at the IPSA Congress in Madrid, 2012)" and has not previously been published.

K. von Beyme, *Klaus von Beyme*, SpringerBriefs on Pioneers in Science and Practice 14,
DOI: 10.1007/978-3-319-01535-4_10, © The Author(s) 2014

post-transition political sphere. I feel emotionally close to him because I once served as a counter-candidate for the presidency of the International Political Science Association in 1982. My victory was due to the cold war: Wiatr was suspected in Moscow of no longer being Marxist, but was not fully accepted in Washington because he claimed to follow a humanist way of Marxist thinking. I was accepted in Moscow as a 'bourgeois scholar' with much interest in Communist countries without being a cold warrior—and accepted in Washington because I was just another liberal positivist who was not suspected of serving as a fellow traveller. This, however, did not prevent certain conservative scholars considering me as a fellow traveller because they were unable to understand why I travelled so frequently to Communist countries. But I was a social scientist and needed the contacts. In history one could work in the archives of Helsinki without ever going to Moscow. When a famous historian was asked by a medievalist how he could write books on Russian history without ever visiting Russia, he answered mischievously: "My dear colleague, have you ever visited the Middle Ages?" A political scientist working on contemporary Russian problems could not get away with such a joke and had to work in Moscow in the Academy of Sciences.

IPSA for me was an exciting experience of policy-making without formal policy advice. Many Westerners resented the fact that we helped to bring into IPSA countries like China and the German Democratic Republic. My argument in a debate with Karl Deutsch, the President of IPSA ahead of my term: "I am in the opposite position to my country up until Brandt: we did not want to recognize something which exists—the GDR. Now we are supposed to recognize something which does not exist: political science in the GDR. I am in favour of doing it nevertheless. Experience teaches that Communist countries stop denouncing a social science as 'bourgeois' and cooperate—as soon as they are accepted on the international level, as they did in psychology and sociology". I was, however, disappointed in the case of the GDR until 1989: unlike Poland or Hungary the jurists monopolized political science, which was never taught in East German universities.

The IPSA crew with which I worked for 12 years in various offices—especially closely with the General Secretary John Trent—was deeply convinced that a kind of convergence would take place in the cold war camps. My Soviet friend Georgy Shakhnazarov—against the rules of the international association, permanent vice president and first vice president in my time in office—later was Gorbachev's most important advisor. Theories promoted by the Soviet leader of perestroika, like 'new political thinking in world relations' or theories of 'interdependence', Gorbachev got via Shakhnazarov, who picked them up in international discussions in IPSA.

I am not going to overrate our political influence, but what happened was a kind of 'social-democratization of Communist thinkers and leaders' in the intellectual sphere. This may help to explain why the powerful Soviet empire collapsed without being advised by political science experts to choose the Chinese solution as at Tiennamen Square in Beijing, with hundreds of protesting students killed.

Social scientists are no longer the Hegelian type of philosopher working on a general historical theory with the assumption of a hidden teleological process. Incremental and partial theories are the norm. According to Thomas Kuhn, 95 % of what scientists do is 'cleaning work' rather than grand theory, ordering the chaotic variety of facts into typologies and specialized propositions. Nevertheless, it is normally agreed that political science has three objects:

1. The description of reality—in certain fields only

(constructivists might argue that there is no reality but there are only involved people's vistas of what they take for reality).

2. The forecasting of trends of development.
3. The normative evaluation of what kind of action should be carried out.

The last two objects are relevant to political advice:

- No politician wants only a description of what there is,
- he wants a prognosis about what will probably happen,
- and finally he will want advice on how he should act.

A German study of policy advice found out that about 76 % of German political scientists claimed to have given advice to politicians, 27 % to parliamentarians and 22 % to governments (Klingemann/Falter 1998: 332). The best-known advisers to politicians in Germany are sometimes those political scientists who have the most illustrious reputations and hold the top positions, such as Fritz Scharpf or Renate Mayntz (Klingemann/Falter 1998: 333). But more often this is not the case. If we compare scientists with the top reputations with those writers and scholars who have most influence on public opinion, there is hardly any overlap. Professors frequently invited on to TV talk shows sometimes also have a high scientific reputation, such as the sociologists Ralf Dahrendorf and Ulrich Beck, but sometimes they have only a marginal role in the discipline, like Arnulf Baring (Höfer 2006: 58f).

If three-quarters of German political scientists claim to work in addition as policy advisors, then we have to be critical. Maybe this is true if every talk with the deputy from one's favoured party is counted, or the engagement of scholars in local politics in cities where they live is taken as deserving the title 'political advice'. Political Science in some respects is not the preferred source for those professors who advise governments:

- Rarely are political scientists recruited—more frequently *jurists* do the job.
- With the growing globalization and economization of society, the *economists* do a lot of counselling.
- In Germany each ministry has an advisory group.
- In the first economic slump we discovered that the great economic crisis—one of the main reasons for the rise of the Nazis—had been handled almost without any scholarly advice. We decided that this should never happen again, so we copied the *American Council of Economic Advisors*.

Lijphart (1977: 223) the well-known Dutch-American political scientist, a kind of 'commis voyageur' in matters of consociationalism from Malaysia to South America, once complained that:

- political science only waits, and remains a spectator,
- whereas the economist intervenes.

This is an exaggeration as so many bons mots. Economic science, however, is a trendsetter for all the social sciences in concentrating on rational choice or on other mathematical procedures. Only the subfield of 'economic policy' still works on prognostic and normative evaluation in a narrative way which can be understood by politicians. Social sciences tend to become politically irrelevant, the more sophisticated they become in applied mathematics.

My personal experience with advising politicians makes me rather sceptical. When I was invited for the first time by a federal chancellor, at that time Helmut Schmidt, for a talk with him on educational policies he said, "My office does invite political advice". I had to correct him, saying that I was glad that he did not care for minor details in his office because, as the ancient Romans already knew, *minima praetor non curat* (modernized version: 'rulers don't care for things of minor importance'). But I had to explain that his office had just asked me to write a paper on elite recruitment and the possible consequences for a change in the electoral law. Schmidt was the typical 'operator'—efficient and opposed to time-wasting. This type of politician is impatient with the complicated deductions of advisers, and therefore turns to decisionism. A German sociologist, Krauch (1972: 41), who worked for many ministries, offered in his book on *Computer Democracy* a caricature of social scientists who sell their knowledge to politicians. Most of them suffer the fate of 'Mr K.' in Franz Kafka's 'Castle': they don't get beyond the antechamber. The only successful advisor he met changed his attitude: Krauch's friend became a collaborator of the secret power in Kafka's castle. He terminated discussions with colleagues and teaching because he had to work for the planning staff of the Federal Chancellor.

Democracies rarely hire political philosophers. Only autocracies made it possible for even great political thinkers and philosophers from Leibniz to Bentham to offer their advice, even to an autocratic ruler like the Russian Czar, something that normally only adventurers and alchemists did, from Casanova to Cagliostro. It was a time when advice to politicians was informal and without problems, such as Mably and Rousseau advising Wiełhorski on the Polish constitution in the eighteenth century. It was a time when political theory still believed in the *grand législateur* who creates great laws, like Moses or Solon. The Solons of today, however, are craftsmen in legislative details.

The comparison of the political scientist with Mr K. in Kafka's 'Castle' was certainly a caricature. Most social scientists are flattered when they are occasionally asked for advice, but rarely try to push their advisory role. Normally they keep to the motto of Prussian aristocrats: "Gehe nie zu Deinem Fürst, wenn Du nicht gerufen wirst" "Never try to meet your king, unless you are invited by him".

This caricature of the self-appointed advisor was maybe accurate for the last time in the era of the planning euphoria in the early 1970s under Willy Brandt, Bruno Kreisky or Olaf Palme, who closely cooperated. The neo-liberal wave of decentralization and deregulation since the 1990s seems to have diminished the need for political advice by social scientists. Times are past when a kind of 'Beiratsphilosophy'—a 'philosophy of advising bodies', as the Austrians called it—was widespread among scholars who saw society in a mechanistic way, like a cybernetic machine. The 'pirate parties' in Sweden and Germany have recently promoted new forms of advice to politicians via the *Internet*, sometimes in creative proposals for legislation, sometimes only in a kind of 'shitstorm' against what they dislike in official politics. The paradigms of political advice have changed very rapidly over time: the planning devices of the 1970s were replaced by theories of steering in the 1980s. In the 1990s a new paradigm was spreading: *governance instead of government,* global regulation, decentralization and emphasis on *implementation* rather than on big decisions and state interventions.

The more social subsystems like science and politics drifted apart, the less political advice was implemented. Sometimes we discovered only later that the reluctant politicians were anticipating future developments more accurately than political scientists. Let me mention just two examples from my personal experience.

1. When Spain turned to democracy after Franco's death, some German political scientists who had worked on Spain were invited to committees on constitution-making in Madrid. We recommended the German system of federalism in order to mitigate ethnic conflict. Spain preferred, however, to offer one pre-autonomy regulation after another to those areas which were most violent, like the Basque country and Catalonia. The German scholars were deeply disappointed. But now, looking back, I understand the Spanish asymmetric and unsystematic approach: it helped to keep the country together and did not invite further possibilities of blocking central decisions, as the German federal system would probably have done.

German lawyers were also invited. They were more successful in their field when they advocated the German constitutional court. Today Spain is following the German model very closely and with much success.

2. Let me end with a second example. The German candidate for the SPD chancellorship invited me to give advice on the coordination of elections for the parliaments in the provinces or sub-units (*Länder*). I began working with the popular opinion the politicians wanted to hear from me: we should harmonize the election schedule in order to avoid permanent mid-term elections every year. In the end, I felt unable to offer what the politicians wanted: I remained reluctant to streamline the elections because this would have meant further centralization, cutting autonomy in constitution-making by abolishing the dissolution of Länder parliaments. To give you an illustration for my final hypothesis in my working paper: after the reunification of Germany in 1990, all the East German Länder had the same schedule for elections. After a few years they drifted apart, because of parliamentary dissolutions—which are normal in a parliamentary system.

10.2 The Place of Scientific Advice in the Decision-making System

What is the conclusion we have to draw from my personal experience? Advice by scholars is frustrating because politicians have to follow a different logical device. This has been called the 'maximum feasible misunderstanding' (Moynihan 1969) between the two alien tribes: scholars and politicians.

- Scholars want durable solutions.
- Politicians think of their next re-election.
- Science has a code oriented towards true or false.
- Economics differentiates between efficient and not efficient.
- The code of law is legal or illegal.
- Ethics thinks in terms of morally good or bad.
- The legal code is however the only binding one. This is on the one hand an advantage,
- but on the other hand it makes juridical thinking inflexible.
- Morality is also rigid and does not like compromise.
- So the politicians—and in some respects the economists—are the only ones to look for compromises in order to avoid "maximum feasible misunderstanding".
- On the one hand the subsystems of politics, economics and science are drifting more and more apart.
- These subsystems try to compensate the distance by interpenetrating each other.
- The more the systems are blurring, the more 'Codes of conduct' have to be promoted, in order to prevent a complete mixture between advising scientists and the lobbyists of interest groups (Althaus/Meier 2004: 238ff).

The frustrations of advising scientists are different at various levels:

1. There is least frustration in *routine decisions*. The 32nd amendment to the German 'Social Security Law' was accompanied by political advice, mostly from economists. These belong to what Americans call the 'iron triangle of parliamentarians, bureaucrats, and interest group representatives'. Sometimes they manage to extend it to an 'iron quadrangle'. In the media they are dubbed the 'social security men'. Oddly enough there are few women among them. A dozen professors have monopolized policy advice in certain highly complicated fields where even most of the deputies in the respective parliamentary committee have insufficient knowledge of the consequences of their decisions.
2. The influence of policy advisors on a *middle level of reform* is different according to the policy field. In a study on legislation from 1949 to 1994, this author found that most of the scholarly interventions in key decisions took place in legal policy (42.3 %), construction and housing (15.2 %), and environmental policy (14.6 %) (von Beyme 1998: 25). Sometimes new key decisions are created with the help of advising scholars, such as the 'Federal law on emissions' of 1974. The interest groups remained largely silent because they did

not anticipate which impact the mostly symbolic law of environmental protection might have on their group interests. One year later a decree had to clarify the specified levels of permitted emissions. All of a sudden the scholars were less important and the interest group experts were active in political advice. The more specialized a field, however, the greater the danger that boundaries between scholars and interest group experts are blurred. There are also of course professors everywhere who permanently work for certain interests. They claim scientific independence, but intellectually become agents of an interest group. A notable example in Germany is the lawyer Hans-Herbert von Arnim. When party finances, the fight against corruption, the salaries of politicians are discussed no talk show can do without him. His books figure on lists of best-sellers. He claims to serve the 'common good'. But many colleagues resent the fact that his empirical work was financed by the 'Federation of Taxpayers', for which he had worked for 10 years before entering the university. As a taxpayer I share his concern for lowering party subventions or salaries for politicians, but as a scholar I may have different ideas about his function in a political system. If we compare the salaries of politicians with what economic leaders receive, politicians are even underpaid. As former German minister Norbert Blüm once put it so nicely. "I would serve for less money, but what I do is more important than what second-rate soccer players do who get twice as much money".

There are sometimes quarrels between different experts. They occur quite frequently between lawyers who ask 'what is legal?' and political scientists who ask 'what is feasible?' They are unproblematic as long as the debates are in public. The more they take place in public, the less important is their contribution to the final political decision. This applies to public hearings in parliamentary committees, a possibility which many countries have copied from the American Congress. Quite frequently they serve as a show which itself serves to legitimate a decision which the iron triangles have reached in a secret meeting of the committee beforehand. Some parliaments have tried to overcome this kind of window dressing. They have also tried to build up a planning system for the deputies. Nowhere has this really happened. Even in the USA Congress tried to limit the President's advisory committees, without success (Wolanin 1975: 198). In America it has been calculated that about 20,000 scholars work in public and semi-public advisory institutions. This inflation has caused a demand for the control of what they are doing—as a kind of protection for possible consumers of their advice (Holt/Turner 1974).

3. The experiences of frustration apply especially to *great structural reforms* such as the electoral system or federalism. There are too many veto groups with vested interests to follow the scientific logic of advisors. Recent theory-building in political reform concentrates on two explanations of why reforms fail:

   (a) Reforms are subject to a first paradigm which states that historical heritages shape decisions and are path-dependent. Previous structural decisions cannot be revised overnight. To offer an example: all western democracies—and even some consolidating new democracies in the East—have to reshuffle their social security systems. Comparative scholars sometimes advise they should copy the United States or the Scandinavian countries. Examination in detail shows that it is almost impossible to transfer foreign models to different political cultures. Conservative (including Christian democratic) liberal and social democratic systems of social security cannot be introduced at will. Even Marxist thinkers who tended to believe dogmatically in a unilinear process of history, like Rosa Luxemburg, once recognized that "you cannot pick up reform or revolution like hot or cold sausages at the buffet of history". 'Families of nations', 'families of taxation', 'families of social security' have been discovered since social scientists learned the lesson that history and historically grown institutions matter.

(b) The second paradigm which rules the discussion on reform policies advised by scholars is Tsebelis's theory of *veto groups*. Among the typical veto groups figure second chambers in parliament, federalist divisions of power, constitutional courts, and the power of presidents in presidential or semi-presidential systems, such as in France or in Poland under Wałesa. Advice on the best solution in an abstract way today is not enough. We need special knowledge of how to find the veto points. One kind of veto group in policy advice is the *Constitutional Court*. In Germany it tends to strengthen scientific advice. Half of the judges have been university professors. Their decisions have many unpredictable consequences. The constitutional lawyers are no experts either. But they hide their responsibility behind formulas such as "the legislator has to consider the state of scientific knowledge" or even the "state of technological knowledge" (BVerfGE 49: 90). Apparently only specialists are capable of knowing this state of scientific knowledge and politicians become dependent on their expertise. Sometimes the lawyers, as in the case of a "law for a statistical survey of the whole population" (BVerfGE 65: 55), have turned down the considerations of the ecologists and protectors of the secrecy of personal data by ruling that the 'state of the art' may one day make it sufficient to interview only a representative sample. But the 'state of the art' so far does not allow the possibility of a federal census of the whole country to be excluded. It was already doubtful that this was correct: on the one hand representative samples are more accurate, whereas an overall survey produces at least 5 % of defectors who do not answer for political reasons. On the other hand, overall surveys are extremely expensive and may serve the interests of a huge interview industry.

Innovative political decisions are normally based on what Paul Lazarsfeld once called 'half-knowledge'. Politicians have a lot of informed guesswork at their disposition, but they cannot wait until we know the future completely. Politicians are like doctors who are trying to cure cancer, though we still do not know exactly how it develops. Why do 30 % of non-smokers die of lung cancer, and why do many heavy smokers never get cancer? Is it genetic disposition only? Another veto point is the opposition in parliament. The opposition is equipped with counter-experts who challenge the truth of the governmental advisors. In a polarized system the parties in power normally select a scholar close to their party. Germany is unique in a kind of bridging institutions: party foundations, funded by the federal government, organize meetings between politicians and scholars. Critics resent this kind of etatism of the parties. The Green party was particularly acute in these matters. It was, however, invited to collusion or corruption by being offered funds for the Heinrich Böll foundation, which functions like other similar party institutions, another proof of the hypothesis that it is difficult to leave a historical path in a country, even for fundamentally oppositing parties. Even the former GDR communists in the PDS (now *Die Linke*) are being integrated into our system via their party's *Rosa Luxemburg Foundation* and gladly accept money from a 'capitalist state'. I once had personal experience of this in my country. Papers on the reform of industrial relations were prepared by the party foundation of the CDU. A former assistant of mine worked in the group, but for personal reasons was unable to finish his paper in time. He asked the decision-makers to accept me as a substitute. They refused a Social Democrat. Finally I wrote the paper under the name of my former assistant—I got the money from my proxy, he got the honour.

Does this mean that the advisors are always party men? According to the Common Procedure of all German Federal Ministries there is a paragraph (I § 61): "Experts should be carefully selected. Decisive factors are acknowledged competence and complete independence in relation to interested groups'. This means, however, that 'reputation' is taken for 'truth', since there is no agreement among the experts on experts. The circle of experts eligible for being selected is restricted even by objective criteria: in the USA it has been calculated that 25 % of relevant papers are published by 2 % of all scholars in the field and 50 % of papers by only 10 % of them (Lohmar 1967: 43). This image of the excellence of the expert is rather elitist and no longer feasible in a democracy. Political advice is democratized for two reasons:

- The experts are no longer lone wolves, the advisers need advice and help from hundreds of assistants.
- The politicians who receive the advice can use it only when they also have hundreds of little experts in their staffs who translate the scientific papers into politically feasible propositions. Political advisors have modernized: they have developed from the role of a 'court clown' to a highly technocratic 'think tank' (Böhret in: Fisch/Rudloff 2004: 379). In the United States alone about 1,500 think tanks have been counted (Althaus/Meier 2004: 49).

If young people study political science they must think of their future profession. In my institute I always ask the students on the introductory course about their visions of future work:

- 60 % aim for the media,
- 20 % see interest groups, parties and parliaments as a source of employment,
- 10 % aim for international organizations and the diplomatic service.

I am old enough to know that this self-image of students does not always work out:

- roughly 20 % get into the media,
- many of them end up in private enterprise,
- 10 % will teach 'civics' in schools,
- a small minority will fail because of over-production in the field and drive taxis or run a gas station (this happens more frequently among American PhDs than Central European ones).

Nevertheless, it is noteworthy that, judging by their professional desires, many students of political science are prepared to work in the field of 'political advice' in political or international institutions.

10.3 The Predicament of Policy Advisors: Incomplete Knowledge in Transformation Processes

The greater a social event, the less we know how to handle the crisis. Knowledge is especially incomplete in times of transformation, as in the case of 1989 when Communism collapsed. Experts from abroad lost their expertise on the *arcana imperii* of Communism overnight. They failed to predict the collapse, and those who predicted some kind of disintegration offered the wrong reasons: war between China and the Soviet Union (Amalrik), ethnic strife (Carrère d'Encausse). Ethnic strife happened, but only after the Soviet Union proved to be unable to hold the Empire economically together. It is noteworthy that most of the former kremlinologists no longer played an important role in East European studies after the transition to democracy.

- In economics the most abstract and ideological experts like the Chicago boys were invited. They had no knowledge of the country, but were stubbornly convinced their advice was good for Poland, since it was good for Chile. Balcerowicz was even fairly successful. New systems of flat tax seemed to stand a chance and contributed to the growth of budding democracies.
- There was a dynamics of transformation, but hardly a strategy, because there was no experience of how to handle the collapse of two dozen states. Moreover, constitution-building had experts, but there were hardly any experts to give advice on changing two things at the same time: the political and the economic

system. There was hardly any precedent in history for a non-revolutionary change of both systems.

In the international arena Communist leaders sometimes suspected that the American secret service had a master plan for bringing the Socialist countries under Western rule. Nothing like that came to light. Kohl managed reunification in the way Machiavelli described successful politicians:

- Kohl had *virtù* (virtue). Germany had a chancellor who was an old-fashioned patriot. There was more of this in Poland, but in West Germany it was rare: as a historian and political scientist, a graduate of my own Institute in Heidelberg, Kohl wanted reunification and was ready to risk something for the allegedly 'flourishing landscapes' he hoped to create. In the international field he had *fortuna* (good luck), because of the crisis in the Soviet Union. Incredible things happened: Germany was reunified. Nobody wanted it—with one exception, President George Bush senior. Mitterand and Thatcher, our closest allies, grudgingly accepted it, because they did not want the European Community to pay for the reconstruction of the GDR.
- Kohl internally created a kind of 'necessità' (necessity). This meant for our neighbours and West Germany: we have to accede to German reunification, since otherwise the support of a democratic GDR will be costly and West Germany would have very limited influence. As the liberal leader Lambsdorff put it, "we cannot allow the East Germans to play a little bit of Communism on our expenses". Reunification was systematically miscalculated. Experts warned that reunification would cost 100 billion euros. The truth was that the costs amounted to 100 billion euros for Europe every year!

What happened to the former experts in transition countries?

1. The former experts in the respective countries who were not simply teaching Marxism-Leninism—and losing their expertise overnight—were overburdened. My colleagues from the Moscow Academy Institute of State and Law under Gorbachev and Yeltsin complained that continuous work was impossible. Every day there was an urgent demand from the President's office:

- one day, how does the French semi-presidential system function?
- next day, it was American federalism or the German constitutional court or Germany's industrial relations;
- another day it was a comparison of West European electoral laws or a study of the Swedish ombudsman. There was a touching scene when one of Yeltsin's great enemies, Chasbulatov (1992), taught a constitutional committee at the Duma about the differences between parliamentary and presidential systems.

Gorbachev was impatient and complained in his famous speech that the advisors did not offer good advice, but 'preferred to write papers in the style of toasts to good health'. But how could it have been otherwise, since for 70 years the experts had been fettered by ideology and controlled by the party? Even

Gorbachev, in spite of his new thrust, stuck to the typical communist manner of explaining failure:

- no *structural analysis*,
- but rather a search for *motivational failures* and culprits among defectors, defeatists and traitors.

In times of transformation linear causality sometimes is not applicable. Chaos theories and theories of catastrophes might be more appropriate. But they have been used only in limited fields, such as environmental policy (Böhret 1993).

Germany underwent a partial transformation—the transformation of the former GDR. The government had incomplete knowledge about the state of the art in East German society and incomplete experience of the comparative politics of transition to democracy. Therefore, from 1990 to 1994, the federal government financed a 'Committee on Social and Political Change' (KSPW). I was a member of its steering committee. We produced a lot of good papers and recommendations. Soon we were allowed to publish them, which demonstrated to us that no 'secret classified knowledge' was suspected. Our work had some scientific value but the government soon lost interest in so much 'specialized knowledge'. The institution, welcomed by the media to great acclaim in 1990, petered out, almost unnoticed by the public, in 1994.

Sometimes Western scholars were asked for advice. Could we be sure that the best Western institutions would function under Russian conditions? The German politicians of the late 1940s sometimes followed their intuition. Ludwig Erhard, the famous father of the economic miracle, leapt into monetary reform. All the international advisors thought that this was premature. He did it, and he was successful. Theodor Heuss, who was to be elected as the first German federal president, was frequently quoted in parliamentary debates for having said in May 1949: 'Whether the constitutional lawyers will be satisfied with our decisions will interest us later when they write their commentaries. The experts will criticize us because they are pedants of perfection on paper—whereas we are actors in a political process.' (12th BT 30. 6.1994: 20962 B).

10.4 Conclusions: The Future of Policy Advice in Democracies

In spite of the above-mentioned predicaments, the future of policy advice is not bleak. There are increasing functions for policy advisors:

1. Scholarly advice serves as an *early warning system*. There are hundreds of councils trying to anticipate problems which have to be regulated.
2. Advisors sometimes serve as *mediators in a conflict between state agencies and social organizations* and interest groups.
3. Advising bodies serve as a *control for decisions* which have already been made. The problem of incomplete knowledge is met by new institutions of evaluation. Our incomplete knowledge about the subject we regulate leads to new forms of

policy advice ex post facto. Evaluation involves scholars. Many laws contain already special restrictive clauses: there is increasingly *sunset legislation*. Laws are valid only for a short time, or *laws require a report by a committee of advisors* after 2, 3 or 5 years.

On the other hand, we have to be prepared that not everything we know and that political science recommends will be implemented. Even if politicians have full knowledge of how to deal with a question, they sometimes ignore that advice for a number of extra-scientific reasons:

1. A decision is not accepted by the majority of the people. Lawyers sometimes hide their misgivings behind public opinion polls. A deputy in a debate on liberalizing the prescriptions on homosexuality quoted a famous professor of penal law and argued: "law does not have the function of serving as a torch-bearer of the sexual revolution" (5th BT. 9. 5. 1959: 12832 D). Sometimes unpopular decisions among parliamentarians are left to a referendum: the Danes declined in 1968 to lower the right to vote to the age of eighteen, the Swedes once declined to switch from driving on the left to driving on the right. After a couple of years, the parliaments decided and the people accepted it.
2. *Ethics commissions* reject a decision, such as liberalizing genetic technology.
3. A decision is necessary, but there is no financial or administrative possibility of implementing it. Some parliamentary laws can never be enforced. The German Greens managed to get a law against 'rape in matrimonial relations' passed. The rest of parliament consented, well knowing that there was hardly any possibility of controlling offenders. *Symbolic politics* can nevertheless be effective, because furious husbands may take the law as a reminder that their behaviour is unlawful.
4. A decision might be rational, but the deputies as well as the politicians are not aware of certain *secondary consequences*. This applies to many decisions in advanced technology or genetic technology. A recent case in Japan and Germany was the sudden decision to close down all the atomic power stations after the accident at Fukushima. In both countries, many awkward compromises had to be introduced after this splendid all-embracing reform—so much hailed by the international media. We know from experience in policy advice that good scholars are not necessarily good advisers (Leitlinien 2010: 12ff). In addition to good ideas, the advisers need a lot of *extra-scientific sensibility*: a sense for good timing of advice and a good understanding of the *logic of the political process*. Sensibility for the fact that not only scholars but also politicians claim a right to normative explications of the advice they receive. Normative recommendations should be clearly distinguished from scientific insights.

There is no organization of political advice which can be generalized for all cases, but there are certain principles such as:

- distance,
- plurality,
- transparency,
- publicity.

There are various types of advisory bodies, such as scholars or groups of scholars who give advice to one ministry or to one party leader, councils of experts in certain fields, and *ad hoc commissions* for executive or parliamentary commissions of enquiry. Distance and plurality is certainly less developed in advising one ministry than in parliamentary commissions of enquiry as a whole. Transparency is better in commissions of enquiry and publicity is more often abused in bilateral relations between scholars and politicians.

My final conclusion is this. The four principles, valid for most systems of parliamentary democracy and formulated for German political science by the academies of sciences, lead to the insight that a technocratic state in which politicians and experts form a kind of estate of 'philosopher kings' is a dream, and not even a desirable one!

References

Althaus, Marco; Meier, Dominik (Eds.), 2004: *Politikberatung: Praxis und Grenzen* (Münster: LIT Verlag).

Beyme, Klaus von, 1998: *The Legislator: German Parliament as a Centre of Decision-Making* (Aldershot: Ashgate).

Beyme, Klaus von, 2002: *Politische Theorien im Zeitalter der Ideologien. 1789–1945* (Wiesbaden: Westdeutscher Verlag).

Böhret, Carl, 1993: *Funktionaler Staat* (Frankfurt: Lang).

Chasbulatov, R. J., 1992: "Kakaja vlast' nuzhna Rossii?", in: *Socis*, 11: 18–31.

Dagger, S. et al., 2004: *Politikberatung in Deutschland* (Wiesbaden: VS Verlag).

Dogan, Mattei; Pahre, Robert, 1990: *Creative Marginality: Innovation at the Intersections of Social Sciences* (Boulder: Westview Press).

Fisch, Stefan; Rudloff, Wilfried (Eds.), 2004: *Experten und Politik. Wissenschaftliche Politikberatung in geschichtlicher Perspektive* (Berlin: Duncker & Humblot).

Höfer, Max A., 2006: "Welches sind die deutungsmächtigsten deutschsprachigen Zeitgenossen?", in: *Cicero*, 4: 58–63.

Holt, R. T.; Turner J. E., 1974: "The Scholar as Artist", in: *Policy Science*, 257–270.

Klingemann, Hans-Dieter; Falter Jürgen W., 1998: "Die deutsche Politikwissenschaft im Urteil der Fachverbreter. Erste Ergebnisse einer Umfrage", in: Greven, Michael (Ed.): *Demokratie – eine Kultur des Westens? 2. Wissenschaftlicher Kongress der DVPW* (Opladen: Leske & Budrich): 306–341.

Krauch, Helmut, 1972: *Computer-Demokratie* (Düsseldorf: VDI-Verlag).

Leitlinien Politikberatung, 2010: (Berlin: Berlin-Brandenburgische Akademie der Wissenschaften).

Lijphart, Arend, 1977: *Democracies in Plural Societies* (New Haven: Yale University Press).

Lohmar, Ulrich, 1967: *Wissenschaftsförderung und Politik-Beratung* (Gütersloh: Bertelsmann).

Marin, Bernd, 1985: "What Is Half-Knowledge? Sufficient for – and When?", in: *Knowledge: Creation, Diffusion, Utilization*, 3: 43–60.

Moynihan, Daniel P., 1969: *Maximum Feasible Misunderstanding* (New York: Free Press).

Wolanin, Th. R., 1975: *Presidential Advisory Committees* (Madison: University of Wisconsin Press).

Chapter 11
The Concept of Political Class: A New Dimension of Research on Elites?

11.1 Introduction[1]

The term 'political class' on the European continent became a populist battle cry to express 'Politikverdrossenheit'. The author of this contribution tries to assess the usefulness of the concept by empirical analysis. The decision-making aspect continues to be treated under the term 'political elite'. Political class makes sense only in those aspects by which the ruling groups stabilise their organisations (the party state) and their income as a group. The tendency to lose ground is, however, compensated by increasing responsiveness. Political class as a notion is wider than the term elite because it includes also the backbenchers. On the other hand it is narrower because the political elite comprises actors beyond government and parliament, such as administrators, media agents and interest group leaders.

An examination of classical texts (Mosca 1953; Pareto/Vilfredo 1916, 1964, vol. 2) makes it clear that meaningful analysis of the term 'political class' is possible only under two conditions:

1. Modern analysis has had to accept the *differentiation of spheres of society.* Political class is not—as in a plutocracy or in the image of state monopoly capitalism—to be identified with the capitalist class. Though the political class tends to become increasingly similar to the economic elite in developing entrepreneurial aspects, it is functionally differentiated from economic leadership. Max Weber was more realistic in this analysis than the Italian school from Mosca to Michels.
2. Political class as a concept for modern analysis has to focus on the *party organisation* aspect. Modern parties should not be analysed under the label of 'conspiracy'—as if a ruling elite used party or *caucus*, or as Ostrogorski and Weber

[1] This text was first published by Klaus von Beyme as "The Concept of Political Class: A New Dimension of Research on Elites?" In: Carlo Mongardi (ed.): *Gaetano Mosca, scienza politica e regime rappresentativo nell'età contemporanea.* Rome, Bulzoni, 1995: 153–177. The permission to republish this text was granted by Carlo Mongardi on 3 June 2013.

K. von Beyme, *Klaus von Beyme*, SpringerBriefs on Pioneers in Science and Practice 14, DOI: 10.1007/978-3-319-01535-4_11, © The Author(s) 2014

prefer to argue—as an instrument to subjugate the ignorant masses. Whereas Mosca and Pareto neglected the party aspect, Michels overemphasised it. The bogey man 'party' was analysed -in contrast to Weber—out of proportion to other fields of political relevance. If political class is identified with some underlying social stratum—plutocracy or some other group—the question of coherence is shifted to another level. If, however, the parties and other institutions are recognised as the essence of the rule of governing class, we have to explain how parties opposed to each other in government and opposition can be considered as parts of a single ruling class.

Modern systems theory developed the idea that since the French Revolution politics was split into ideological camps. There were always two aspects of the political situation, institutionalised in government and opposition. The secondary codes of the parties are not social reality, but there are ideological perceptions of political realities and therefore they are mutually interdependent. One could try to ontologise these interdependent beliefs of a political class. This would, however, exclude ideological civil war. The idea of a binary code applies only to systems which are no longer endangered by revolutions and *coups d'etat,* in which the opposition regularly and grudgingly accepts the victory of the government party. This type of elite has been dubbed a 'consensual unified elite', characterised by the willingness of most or all influential persons to sacrifice partisan interests in order to preserve existing institutions where the two conflict (Highly et al. 1976). The older elite theorists thought of elites as a much more unified group. The fact that most members of a political class in making innovative decisions—not in routine decisions—normally implement only their second choice in order to preserve the rules of the game of the existent order was underrated by both Mosca and Pareto.

Thus the problem of proving the coherence of a political class is less difficult than in the older theories of political class. The task is made unnecessarily difficult if one follows an analyst of Mosca's work such as Meisel (1958) who proposed the 'C-formula' which overrates the coherence of a political class more than Mosca or Pareto themselves ever did. The characteristics of a political class according to Meisel were 'consciousness, coherence and conspiracy'. This last term was watered down in the German translation, authorised by the author, to *Zusammenarbeit,* which would mean only 'cooperation'. The C-formula has been widely attacked as not being in tune with Mosca's and Pareto's intentions. Even so, the formula could still be examined according to the criterion of its applicability to the problem: if we discard the misleading notion of 'conspiracy' the three elements hint at the two levels which have to be used in order to demonstrate the existence of a political class:

- What is the political class?
- What does the political class do?

The first question seems to aim at a concept of *elite,* the second at a concept of *political class.* Research, however, has hardly taken these terminological clarifications into account, and 'elite' is used to cover both aspects.

For the first question the term ‘elite’ has prevailed, for the second question ‘political class’ has been used (Kadushin 1968). This division has not convinced all researchers. Elite structure and policy output have been studied predominantly without the use of the concept ‘political class’. The typology this author would rather postulate is that of a close connection between the concept of political elite and a theory of political action. In America the term ‘elite’ was sometimes considered ‘static’. Behaviouralist authors (Welsh 1979: 17) therefore preferred the term ‘leadership’ for more dynamic investigations. Leadership as a basic concept was less popular in Europe. It was used, however, to explore the relationship between elites and non-elites. A matrix (see Fig. 11.1) could exemplify divergences and overlappings between the two concepts ‘elite’ and ‘political class’. On the X axis, approaches of action theory and structural system’s theory are shown. On the Y axis, the predominant interests of leadership groups are identified. In the second field of the matrix both concepts tend to overlap, and in terms of recruitment and professionalisation both offer an almost identical field of research. The following four differences between elite and political class can be stated:

1. Elite is the broadest idea employed in leadership research. It also comprises economic, cultural and media elites as far as they influence the political process. Only in some respects does the term ‘elite’ seem to be narrower. Political class encompasses all those politicians who participate in the privileges of the elite—even if they have only a minor importance for the decision-making process.

   A political elite is the group which predominantly produces decisions binding for the whole political system—even against resistance from the economic and social subsystems of the community. A political elite has to cooperate with other elite sectors in order to maintain its steering capacity. The political class, on the contrary, can rarely cooperate with other sectors in order to broaden its privilege structure, for it can count only on the passive tolerance of the voters and other elite sectors. This is the main reason why tendencies of ‘losing ground’ (*Verselbständigungstendenzen*) grow in modern democracies. As soon as a new scandal is uncovered, the media elites and the leaders of interest groups tend to blame the political class for its uncontrolled actions. Only the economic elites are less concerned because they know that the rewards and privileges of the political class are still small compared with their own. A political elite acts on behalf of a third party, the public. Its main interest lies in maintaining steering capacity. The political class, on the other hand, is self-referential: it develops interests only on behalf of itself. The political class needs a certain autonomy from its electors. But it compensated for the impression of being aloof from the interests of the public by responsiveness.

   The political class develops its autonomy via decisions on the periphery of the decision-making process (laws on the remuneration of politicians and party finance), and in strengthening its own influence in other spheres of society by penetration of the leadership positions in the public mass media and in the

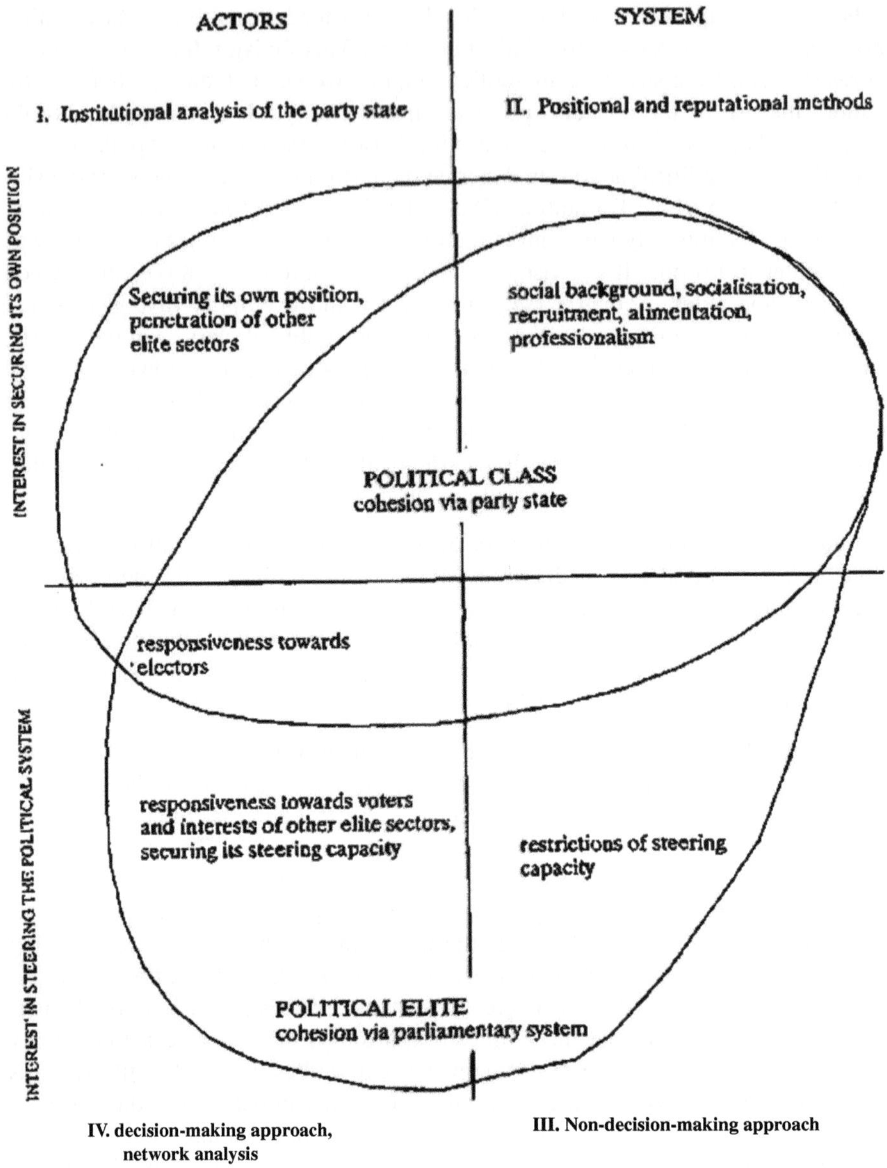

Fig. 11.1 The focus of the concepts 'political class' and political elites

realm of the state economy. There is, however, a personal union between the members of the two groups. But the political class cannot mobilise the voters for a decision on its privileges as the political elite can mobilise voters to make important decisions in various policy arenas.

2. The political class thus also *comprises the backbenchers* who are only marginally involved in the important political decisions. In the elite, on the other hand, backbenchers play a minor role because of the hierarchy in the decision-making process. Although backbenchers play a secondary role in the decision-making process, they are frequently the driving force behind measures to secure the privileges of an almost 'egalitarian' political class. Elites are mainly motivated by power interests, a political class is motivated by interests of economic and social security for itself. The members of a political class therefore resist hierarchies. In parliamentary groups of many continental countries it is still extremely difficult to violate the egalitarian principle so as to secure extra benefits for the leadership groups in the hierarchy of political elites, such as leaders of parliamentary groups or chairpersons of parliamentary committees. The *classe dirigeante* is part of the *classe politica,* but not the other way round. Political elites are identified by network analysis as a centre of decision-making, whereas the political class can be identified by the old-fashioned positional approach. The concept of a continental 'party state' is the main instrument of securing privileges for the political class. Party studies are therefore still more important than the study of elite decisions.
3. The changes in recruitment and socialisation of both leadership groups are mainly a product of unsteered social development. The parties try to steer this process by quota and recruitment guidelines in the constituencies, usually with little success. Even if they were successful the benefits of elite decisions for the voters would count more than improving the social representativeness of the leadership groups. The political class comprises the oppositional parties as long as they do not aim at 'another republic', that is to say, remain loyal to the constitutional system. It secures its cohesion mainly via the party state. Political elites are bound together by the rules of the parliamentary game. The folk-ways of parliamentary government also have, however, an important influence on the cohesion of the political class via a shared ethic of the 'right honourables'. Therefore the tendencies of 'cooperative parliamentarian-ism' and co-government of the opposition have to be studied in continental party states in order to show the growing cohesion of the political class.
4. Research on elite and political class varies in method. The study of the 'party state' was rarely included in examinations of elites (Field I). The *positional method* (Field II) proved to be the most economic type of research. It identified elites with official offices. But decision-making studies show that many important decision makers do not hold formal office. The positional method has been more useful for defining the political class, those who participate in the privileges of government, parliament and top administrators.

The *reputational method* (Field III) completes the finding of the positional approach. Each method works with simplifications. Even the actors, when interviewed, work with simplifications. No interest group can influence every elite member or even bribe it. Nevertheless, they are successful in finding out the important actors with whom they cooperate-also applying a kind of reputational method. The classics like Mosca (1953) were vague in the figures they offered.

Mosca sometimes mentioned ‘a dozen’ and sometimes a ‘hundred’ politicians. Reputational studies tend to be more precise, even if they sometimes tend to be pseudo-exact, when for instance they come to the conclusion that in the USA there are 876, in Australia 746 and in Germany 799 leaders involved in a broader circle (Highly 1991: 35ff). But for the study of the density of networks these figures are important. Network analysis, in a second step, came to inner circles (Australia 11—USA 32). Business and media elites in these core elites were under-represented, and trade union elites proved to be important only in the German case (this was, however, the artefact of the SPD. government then in power). The more network analysis was related to decision-making, the more interest groups formed a part of the core elites. In a pluralist system the core was sometimes found to be ‘hollow’ (Heinz et al. 1990) which tended to distort the reality. Institutional studies therefore had to correct the finding, via a reputational method and by network analysis.

The *decision-making approach* (Field IV) was decisive for the search of the political elite. It was, however, also combined with the reputational approach. Restrictions in the political system, which impede every attempt to steer the whole society, forced students of elites to develop the *non-decision-making approach.* It proved to be important not only to ask ‘what does the political elite do?’ but also what it does not do (Bachrach/Baratz 1970: 44). Leftists preferred this approach. They no longer adhered to a simple Marxist ascendancy theory. But on the question ‘what does the ruling class do when it rules? (Therborn 1980) the answer tended to end up with a truism. Essentially, it reproduces the economic, political and ideological conditions of its domination. There was a hidden teleological view on history: history aims at socialism. The political class, however, seeks to impede this ‘natural development’. The results thus tended to be tautological. It stands to reason that capitalists do not aim at socialism. In times of autopoetic systems theory, we are more prepared to believe that no system tends to develop a new order contrary to the basic rules of its functioning. The non-decision-making approach is relevant only when it is not restricted to great controversies (Sartori 1987: 141–176). It is fruitful, however, only when it transcends descriptive studies of single decisions. In comparisons of a few *routine decisions* no coherent interest of the actors has been found (Laumann/Knoke 1987: 383) Miscalculated strategies and ‘unholy coalitions’ can operate against the well-understood interest of single interests within the status quo. Only comparisons of a sufficiently large number of innovative decisions avoid the fallacy of premature conclusions on non-decisions.

11.2 Empirical Evidence for the Existence of a Politial Class

If we examine again the characteristics of what social class is, and second what it does, we should look first at the tendency towards the social homogenisation of the political class.

11.2.1 Social Homogenisation

In some respects the term political class belongs to a pre-democratic oligarchic society, to a period when a mixture of the old democratic society and the new bourgeoisie ran the state. The homogeneity of this elite did not correspond with the period of latent civil war in the first half of the twentieth century. In developed democracies the distinctions between class and elite interest became more apparent. The period of classical modern times was characterised by a stalemate of *asymmetrical elites:* the bourgeois elites predominating in the state, and the working class elite which was usually confined to a veto role and only occasionally won a hegemonic position, as in Scandinavia. In post-modern times the elite structure developed similarities with the pre-modern conditions of a political class:

1. *Ideological divisions* played a secondary role.
2. The social background of elites became more homogenous. Neither the older upper classes nor the working class were a major ground of recruitment. Middle-class civil servant elites held a strong position, and they were socialised in the feeling that political elites live not only 'for' politics but also 'on' politics. According to empirical studies even the voters have modernised their perceptions. The connection between social background and representatives has weakened, and the majority of voters found the recruitment of deputies from their own social group no longer as important as in former times (Herzog et al. 1990: 56).
3. Education in the pre-democratic era was at the root of the main gap between the elites and the non-elites. The differences have now been levelled down. The functional differentiation of educated elites, however, gave many opportunities to the educated, and politics became rather an unusual field of activity for the cultural elites. In times of rapid démocratisation, after 1945 in the former Fascist countries, in the 1970s in Southern Europe, and recently after 1989 in the ex-socialist countries, the intellectuals played a certain role in the first cadres of democratic elites. But this impact withered away very quickly.
   Political elites today are more educated than in former periods. Mosca and other elitist thinkers had hoped that education would be a condition of elite recruitment. Some of this wishful thinking has been realised in modern democracies. Education for political elites also plays a greater role than in the pre-democratic political class. Homogenisation of the political elite has increased in modern *meritocracy.* The socialisation and education patterns of the political camps, conservative or Christian democratic parties and social democrats, in many respects grew more alike (von Beyme 1974: 51ff).
4. *Professionalism* of political elites is increasing in most countries. One third of parliamentary elites have only had political experiences. The number of years

in a profession are declining in most continental countries (Kaack 1988: 131) and even slowly in Britain. With social democratic governments professionalism appeared to increase more quickly than in times of Christian democratic rule in Germany (Armingeon 1986: 25–40). But professionalisation has many aspects. The indicator of the number of years in professions versus full-time political activities is only a very superficial one. There are others, for instance, the inclination to stay in politics: even after resignation from the highest office the members of political elites remain in the political class in a broader sense. Horizontal mobility has grown—not so much between politics and economics as it exists in the USA, but between political administrative and quasi-political activities in interest groups.

Professionalism can be shown in the career perceptions. Specialisation in one or two policy fields leads to more substantial knowledge than was the case for the older unprofessional elites. The books politicians write may not always meet scientific standards, but they are much more substantial than in the 1950s. The voters in the process of selection of candidates usually prefer politicians who have some standing in a reputed profession. The tendency to increase remuneration for deputies is the most unpopular issue. By way of growing responsiveness, however, professional politicians can compensate for their lack of professional experience: Chancellor Kohl, a professional politician since obtaining his doctorate in political science and without substantial professional experience, shows that his organisational talent, the main virtue of which was conflict management both within his party, CDU, and with the Bavarian CSU as well as the FDP, was more important than professional experience and specialist knowledge.

If parties fail on many counts, such as interest representation and concept formation, they are certainly more efficient than ever in performing the recruitment function. Non-party politicians have hardly a chance. Even in Ireland they declined where they were most common after 1945. The German Federal Constitutional Court intervened on behalf of a non-party candidate. Such action did not stop the trend of professionalisation of party politicians. A double professionalisation is required in order to win a seat: in politics and in party politics. Nonprofessional politicians are ridiculed as dilettantes. This was quite obvious when the new elites from the former GDR entered the Bundestag and when the state parliaments were constituted in East Germany. The West German press was full of disdain for the 'lay actors' on the Eastern political scene.

The professionalism of the political class varies according to different types such as (Panebianco 1982):

- managers
- notables
- representative bureaucrats
- staff professionals
- crypto-professionals (in parastate institutions)
- semi-professionals (parliamentary staff)

In northern Europe some of these types are rarer than in the south. Where there is an ideological tradition of modernising socialist and communist parties, the 'administrative intellectual' plays a role (Fraser 1976) which is rare in the pragmatic Social Democratic and Labour parties of the north. But even without differentiation of the various kinds of 'organisation man', in most countries professionalisation is on the increase and creates various blends of political leaders and professionals which helps to forge the coherence and success of the political class.

5. With professionalisation, types of income are becoming more similar for the political class. In times of a strong influence of Marxian thought on the theory of stratification there was a great reluctance to identify elite with class, or to accept the concept of a 'political class'. There is, however, little obstacle to this identification in the original Marxian concept of class. Towards the end of *Das Kapital,* in his analysis of class, Marx classified classes on the basis of a "Dieselbigkeit der Revenuen ûnd Revenuequellen" (identity of revenues and sources of income) (Marx 1968: 893) Max Weber distinguished *Erwerbsklassen* (according to the criterion of income) and *Besitzklassen* (according to the criterion of possessions). With the growing welfare state the transfers have increasing importance. *Versorgungsklassen* (classes according to the extent they receive public transfer payments) have been added. All these classifications, according to revenue sources, are typical for this group, especially with growing professionalisation. Political elites combine increasing income—they are the only group which fixes its own share of the state budget by simple parliamentary decision—with prospects of good pensions after leaving politics and with many related privileges. Since at least a third of the deputies are civil servants, they are the only group which makes sure that there is no career break during their active involvement in politics. Civil servants can not only rejoin their former service but in many countries resume office under conditions of a normal career advancement.
6. There are some differences in the *means of information* available to the party segments. Government party members have easier access to information from the ministerial bureaucracies of 'their' respective party members in ministerial offices. But in Germany even the Green party which at one time considered itself as a kind of 'fundamental opposition' had numerous contacts with civil servants (Puhe/Würzberg 1989: 80). The real difference lies rather in the amount of substantial information the deputies of government parties or oppositional parties obtain from these contacts.
7. *Lifestyles* of the political class are becoming more alike. The differences in lifestyles in the bourgeois and social democratic parties vary in Western democracies. In Britain they tend to be substantial, in Germany they are minimal. The living conditions of a small unattractive capital such as Bonn have not facilitated the differentiation of lifestyles. The members of the political class until 1991 rather cultivated their regional images, and the houses and the middle class lifestyle of their families are on the whole comparable across the parties.

11.2.2 Political Class: What It Does: Tendencies of Separation from the Non-Elites

Political class as a concept was coined in a pre-democratic era. Parliamentary government was dominated by highly elitist cliques without clearly structured parties. The *trasformismo* period facilitated the Mosca type of criticism and pessimism. With democratisation of the political systems the elites have become more responsive. This might contradict the possibility of an increasing separation of the political class from the non-elites or voters, but it does not. The concepts of elites or political class were deeply shaped by the experience that the normal citizen does not much care for politics. Michels (1989: 42ff) went furthest in arguing that there was a *Führungsbedürfnis* (a need to be led) by the masses. The period of classical modernity in political theory, however, was still on the mobilising side: the inclusion of more and more citizens was the aim. Social movements no longer aimed at a complete takeover of power, as they had in the nineteenth century, but they were prepared to win their due share of power. Politics became less compartmentalised. Competition for the coats of the new middle classes became fiercer. Responsiveness towards the needs of smaller groups grew. Nevertheless tendencies of organisational separation from the non-elites were growing.

1. Participation has shifted in a post-modern society. *Political entrepreneurs* are becoming more important. The 'resources approach' to social movements does not take ideologies and movements for granted, but asks how single success stories of organisation can be explained among thousands of failures to organise a movement (McCarthy/Zald 1977: 1212ff). The most important resources which decide on success or failure are available leaders, organisational know-how and fund-raising capacities:

2. Even the counter-elites in the new social movements follow the same pattern of development. *Cycles of overproduction of intelligentsia* create the necessary counter-elites. It is not by chance that most of these German leaders were unemployed teachers or social workers and other intellectuals who were 'blocked spiralists'. The 'new class' of cultural elites is partly blocked by the old party elites. The new ones tend to be either *Aufsteiger* or *Aussteiger,* those who adapt to the old machines but propagate ideas of *new politics;* and those who leave the traditional political arena of the established parties and enter new progressive movements (Feist/Liepelt 1983: 94).

   Theories of the new middle classes, seen in the light of new cleavages, have been linked to a theory of the 'new class', value structures which, deviating from the predominant paradigm of a growth economy, are used as a criterion for class divisions to replace the old economic characteristics of income, professional activities, and conditions of life. There will be objections to this kind of typology of class divisions, unless these 'classes' prove themselves when

confronted with the organisational needs of movements. There is, for instance, a difference between technocratic specialists running their machines and the new movements which try to protect their clientele against the dominating technostructure and work for their autonomy (Kriesi 1989: 1078ff).
Comparisons of organisational cultures of elites and counter-elites over time show, however, that even the new class cannot prevent being permeated by the organisational imperatives of the political class dominant in the system.
The new counter-elite has been compared in its behaviour to the socialist elites at the end of the nineteenth century. There are superficial similarities: the new oppositions regularly oppose the state budget. In Germany, the new opposition quickly got involved and started a policy of amending laws and budgets. The federalist structure of Germany gave them early access to power. The first Green minister wore his jogging shoes when taking his oath in parliament. The second time he took his oath in the Land of Hesse he came in a dark suit. The third time he wore already a tie. The acceleration of integration is a striking feature of post-modern politics. It is facilitated by elites which come from the same middle class strata as the established elites.
Oddly enough, the anarcho-syndicalist organisation ideology of the new movements strengthens the party bureaucrats. Michels had predicted that the party bureaucrats would be the new oligarchs who run the big machines. In this he was wrong for most parties. Only for the libertarian movements was he correct. Because the deputies were subject to the rotation principle and the party and parliamentary group bureaucrats did not rotate their power in the way Michels had predicted. Michels (1989: 340) had also predicted that the anarchist movements would create new types of oligarchies. He was taking issue with Ostrogorski whose remedy was the battle cry for caucus dictatorship: "à bas les partis, vivent les ligues!"
Michels saw correctly that even loose anarchist leagues would create the functional equivalent of oligarchy in the traditional parties. The room of manoeuvre for prominent Green deputies was kept narrow. They were humiliated at party meetings and subjected to permanent control. The insecurity of parliamentary elites in the name of democracy led to the rule of bureaucratic agents in the central apparatus of the Green Party.

3. Political competition for votes is increasingly *commercialised* by the political class. In the United States there are few campaign activities which are planned by a party as a whole. Individual campaigners may get help from commercial experts and managers. In Europe, where the party and its bureaucratic structure plays a greater role, campaign managers from outside the parties are hardly thinkable. But there are experts within the parties. Even if they are recruited from commercial institutes they are active within the party. Reliance on the expertise of pollsters is quite common, but pollsters close to the centre of power are a new development (Hofmann 1991).

   Campaign managers are increasingly professionals and familiar with the organisational life of the party. They are usually employees of the party or come from some agency which was founded by the party. The commercialisation of party expenditure is usually still restricted to the central party level and has not yet reached the level of the individual deputies to the same extent as in America (Landfried 1994: 297). Since many European systems have still some kind of public service television, the candidates do not have to buy television time as in America. This also limits the trend to commercialisation. But compared with former periods, the political class is also shaped by the dependence on consultants and campaign managers which unifies the working conditions in all the parties and leads them to conclude electoral agreements for a limited inter-party warfare.

4. A notable trend in the development of democracies is the *étatisation of party finances*. Germany was a forerunner and tried to cover the costs of party democracy to excess. The Constitutional Court had to limit this practice to the refunding of costs for electoral campaigns. Many European countries from Sweden to Italy have followed the example. In newer democracies, such as Spain, the proportion of state funds in party finances is even higher than in Germany, where it amounts only to half of revenues in election years (von Beyme 1985: 206). Even France gave up its resistance to public financing of parties, and only Britain still resists the universal trend of étatisation. The Anglo-Saxon tradition preferred to help party organisation in an indirect way by exempting donations to parties partly from the tax bill. Some countries, such as Germany, tried to combine the best of two worlds and went far in both: American tax subsidies as well as direct subsidies to parties from the state budget.
   Public funding of parties leading to the strengthening of a political class is relevant in several ways: on the one hand the parties develop an interest in common. In case of being summoned to the Constitutional Court German party treasurers usually agree in preliminary meetings in advance what they are going to declare and which figures they disclose. Certain clauses in the laws which regulate party finances were virtually written by the cartel of treasurers. Even the party elites had hardly any insight into increasingly complex regulations and the systems of calculation.
   With growing public subsidies, the parties are less dependent on membership subscriptions with growing public subsidies. The independence of the political class from its voters is fostered by this development. The paying morale of the party members is weak and the party organisation is not compelled to enforce the regular contributions. Only few clauses—as in Germany—bind the state subsidies to the level of mobilising efforts of the parties. Only the introduction of some kind of matching funds could stop this process.
   Another aspect of the growing independence of the political class is the development of a 'rifle type' of private donations. Many businesses no longer choose to donate to the party as a whole but select individual deputies and influential policy-makers or party factions. Thus the connection between donations and political favours in exchange is becoming closer, sometimes near to corruption,

as recent financing scandals such as those by the Flick enterprise, have shown. Some observers even spoke of *capitalization* to characterise this process. Where a party foundation receives public funds, as in the Netherlands and Germany, étatisation has another dimension. The new rebel elite of the Greens fought against the privileges of the party foundation. But the watch–dog itself was corrupted when the Constitutional Court decided in 1986 on an action by the Greens against global subsidies to the party foundations, that the complaint was irrelevant because the Green party was not damaged in its rights since it had no foundation. One Green speaker has correctly stated that this judgment sounded like an 'invitation to corruption'.

5. A further trend to create a unified political class is the move to cooperative parliamentarianism and the development of 'co-government' of the opposition. Kirchheimer once criticised the ideologically faceless catch-all parties. Nostalgia for class conflict was not reflected in the party system, even though it dominated the discussion on parties in the late 1960s and 1970s. Even in Italy the semi-fundamental opposition of the Communists normally carried three-quarters of the legislation with the governing Christian Democrats. Even where an opposition would prefer to vote against a government bill it sometimes cannot, because it would be blamed by the government and voters for the failure of the bill, and using obstructive parliamentary tactics—even outside parliament the members of the political class co-operate permanently. Especially in Germany, proportional arrangements for most boards of management and supervisory bodies, even in the media, are ritualised and create permanent contacts among politicians from various parties. This informal co-operation strengthens the coherence of a political class.
6. The policy arenas 'organise' co-operation with the political class in a wider sense. In the United States, researchers long ago discarded the idea of a political class identified with representative functions of politicians. *Iron triangles* in decision-making—in which state officials co-operate with deputies and interest group agents—show that there is not one clearly discernible political class for all issues. Various policy fields involve different actors. The location of a political class is somehow still linked to an old-fashioned concept of the state. It is frequently assumed that the political class 'runs the state'. The division of private and public is maintained but the boundaries are increasingly blurred. This is not only so in corporatist systems, where the state delegates a good deal of public tasks to private organisations. In non-corporatist systems, such as that of the United States, the 'iron triangles' are a functional equivalent of the blurring of borderlines between public and private, and empirical studies found no magnetic centre for the cohesion of policy-making structure. The network is also highly fragile. No government officials, as in corporate systems, try to smooth down the aspirations of individual actors. The elite network has been compared to an elliptical structure lacking a centre. The 'core' has proved to be hollow. In contrast, European researchers and corporatists still tend to locate some state agent as the coordinating actor. 'Political class' in this type of research is restricted to one arena. The

cohesion is upheld by communication: communication among influential takes place predominantly with the elites of adjacent interest groups, and they deal with their allies rather than with their adversaries. (Salisbury 1987: 1217) The theoretical approach to the co-operation has focused either on an *etatist* outlook which still assumes that state elites organise the co-operation in a corporatist system of exchange or else has classified the exchange relations and looked for the special features of political exchange. Generalised political exchange was made independent of the institutional framework of the state.

The more simple exchange systems include:

- market exchange systems,
- conflictual exchange systems,
- representative exchange systems.

The more complex systems include:

- clientelist exchange systems,
- mediatory exchange systems,
- political exchange systems.

Exchanges are called 'political' not because the participants involved are political actors, such as parties or bureaucrats, nor because the resources exchanged are non-economic. The actors are fewer, and actors are less equal than in the economic sphere. The networks are not temporary but have a certain durability. The political dimension implies that the relations are not just *dyadic barter* structures but *triangular* or even more complex exchange networks (Marin 1990: 64). With such a wide concept of exchange relations the term 'political class' would seem to be under-complex. But in an empirical field situation the intermediation of political actors materialises to the extent that the role of elites or a political class is not a mystification.

11.3 Conclusion

Political class in many European countries has been loosely reintroduced at the popular level. Research, however, continues to be conducted under the heading of elites. The old conflict between Mosca and Pareto about the priority and feasibility of their theories can be continued in examining the usefulness of this concept. The easiest way out of this conflict would be to restrict

- *elite* to the *static* aspects. This concept is used when we ask 'who are they'?
- *Political class* is connected to the more dynamic aspects of the problem when we ask the question 'what do they do?'

A similar dichotomy was introduced in the American discussion:

- *elites* have been contrasted with

- *leadership*. The behavioural revolution even tried to substitute the notion of leadership for the concern of elites. Leadership—in terms of action and judgement about the results of action—was more measurable than the 'qualities' of an elite. This approach also appealed to behaviouralists. The study of leadership was a similar move away from formal political positions, as was intended with the introduction of the term 'political class' after Pareto. Elites need not be leaders, and leaders do not necessarily belong to the elite in a formal constitutional sense. Leadership, however, frequently focused in an old fashioned way on important figures in unusual situations. Leadership was also highly important in the field of symbolic politics, especially at times of elections. Party identification and identification of voters with their leading candidates became an important field of electoral research. Only the more psychological approaches sought to determine static properties of leadership. Generally the concept of leadership was used more to explore 'what elites do'.

Political class has the virtue to be open to both:

- a static exploration of who they are,
- a dynamic analysis of what they do.

Objection to the application of the class concept became less important when the research on class lost its Marxist connotations on the European continent. Marx's definition would not preclude speaking of a political class, if it can be shown that it refers to a coherent group living on politics and having the same types of income. Later developments of the class concept—including the *Versorgungsklassen* of modern analysis—suggests that 'political class' is not necessarily a misnomer. In many fields evidence of coherence and cooperation can be shown which allow reference to a political class. The term makes sense in the I context of a contradiction: The more responsive democratic elites have to be, the more they constitute themselves into a separate class in order to meet the demands made on them. The more politics loses ground and ceases to be the steering mechanism of society, the more politicians resemble other elites, especially economic elites. What Max Weber called the *Anstaltsbetrieb* makes political organisation and economic production units increasingly similar.

Older elite theories from Mosca to Pareto underrated institutions. Later elite theories, from Ostrogorski to Michels, overrated the party bureaucrats. The older idea of political class was appropriate for a pre-democratic system in which a highly oligarchic parliamentary elite played an important role and used its organisation as a substitute for political elites is now different. Still, there is no single class which recruits the political class in an organic way. On the contrary, the diversification of recruitment sources is growing. The balance between a bourgeois political class and the working class leadership which Pareto called *demo-plutocracy* is withering away. Social background, education and lifestyles are becoming more alike in all parties, including those of socialists and social democrats.

Political class was intended by Mosca to preclude normative uses, although Pareto thought that this was much better achieved by his term, elite. Political class

	Social structure approach (asking what the ruling minority is)	**A unit of political action (asking what the organised minority does)**
Minority as a loose group	Elite (Pareto) Capitalist Class (some Marxists) Plutocracy, Demo-plutocracy (Pareto) Intellectuals (Gouldner)	political class (Mosca) leadership (Welsh)
Minority organised in an institution	classe dirigente (Pareto) Caucus (Ostrogorski, Weber) stratum of professional politicians (Weber) party bureaucrats (Michels) -managers (Burnham)	power elite (Mills) Non-Decision Veto group (Bacharach, Therborn, Offe)

Fig. 11.2 Classification of elites. *Source* The author

in the recent context of changes does have a certain normative value. It measures the elites according to the yardstick of democratic elites which are only temporarily delegated and are in constant communication with their voters. There are many tendencies in modern democracies which strengthen the separation of a political class from its basis, such as public financing of parties, monopolisation of political activities by the parties, the co-operation of government and opposition. Though scholars will try to avoid this normative dimension, the virtue of journalistic references to the political class is that it reflects where this separation of an organisational class from its basis should be counteracted in the name of democratic values. There are different concepts of elites or political classes. If we try to schematise them in a matrix, some theories conceive the ruling minorities as a social-structural unit, others look rather for a political acting group. If we relate these differences on one axis to two other perceptions—an elite as a group or as a coherent set of people connected with each other through certain institutions, we get the overview given in Fig. 11.2.

The action-oriented approach postulates a more coherent group than the social-structural ideas of an organised minority. In empirical studies, however, the political action centres prove to be rather loosely structured. Sometimes the core proved to be hollow. In theories which try to explore the steering of whole political societies, such as the corporatist approach or the new abstraction of a generalised political exchange, sometimes do not even require participation in elite action in order to count as belonging to relevant leading groups. In these models political class is a mere abstraction including both structural and action-oriented criteria. Work on a theory of the political class has not only to add the tendencies which hint at a growing independence of the political class from their voters. It has to use the 'method of disagreement' as well, and to analyse those factors which show that there is greater connection between elites and masses. Growing issue-orientation of rational voters enforces more responsiveness on the part of the political elites. Paradoxically, however, the manifold processes which show that the gap between the political class and the democratic people is growing is no mere supposition. In functionalist terms one can argue: the more responsive elites have to be, the more necessary a certain autonomy of a political class from those who entrusted it with power becomes, precisely in order to meet the demands of those voters.

References

Armingeon, Klaus, 1986: "Die Bundesregierung zwischen 1949 und 1985", in: *Zeitschrift für Parlamentsfragen*: 25–40.

Beyme, Klaus von, 1985: *Political Parties in Western Democracies* (New York: St. Martin's Press).

Beyme, Klaus von, ^{2}1974: *Die politische Elite in der Bundesrepublik Deutschland* (Munich: Piper).

Laumann, E. O.; Knoke, D., 1987: *The Organizational State: Social Choice in National Policy Domains* (Madison: University Wisconsin Press).

Fraser, J., 1976: *L'intelletuale amministrativo* (Naples: Liguori).

Mosca, Gaetano, ^{4}1953: *Elementi di scienza politica* (Bari: Laterza).

Puhe, H.; Würzberg, H. G., 1989: *Lust und Frust: Das Informationsverhalten des Deutschen Abgeordneten* (Cologne: Informedia).

Heinz, J. P. et al., 1990: "Inner Circles or Hollow Cores: Elite networks in National Policy Systems", in: *Journal of Politics*, 52,1: 356–360.

Herzog, D. et al., 1990: *Abgeordnete und Bürger* (Opladen: Westdeutscher Verlag).

Higley, John et al., 1976: *Elite Structure and Ideology* (Oslo: Universitetsforlat; New York: Columbia University Press).

Higley, John et al., 1991: "Elite Integration in Stable Democracies", in: *European Sociological Review*, 35–53.

Hofmann, G. 1991: "Vom Wahlforscher zum Politikberater", in: *Die Zeit*, 30,2.

McCarthy, J. D.; Zald, M. N., 1977: "Resource Mobilization and Social Movements", in: *American Journal of Sociology*, 84: 1212–1241.

Kaack, H., 1988: "Die soziale Zusammensetzung des Bundestages", in: Thaysen, Uwe et al. (Eds.): *US-Kongress und Deutscher Bundestag* (Opladen: Westdeutscher Verlag).

Kadushin, Ch., 1968: "Power, Influence and Social Circles", in: *American Sociological Review*, 33,1: 685–699.

Kriesi, H., 1989: "New Social Movements and New Classes in the Netzerlands", in: *American Journal of Sociology*, 95: 1078–1116.

Landfried, Christine, ^{2}1994: *Parteifinanzen und politische Macht* (Baden–Baden: Nomos).

Marin, Bernd (Ed.), 1990: *Generalized Political Exchange* (Frankfurt: Campus-Boulder, Westview).

Marx; Engels, 1968: *Marx Engels Werke*, Vol. 25 (Berlin: Dietz).

Meisel, John, 1958: *The Myth of the Ruling Class. Gaetano Mosca and the Elite* (Ann Arbor: University of Michigan Press).

Murphy, W., 1990: "Politischer Protest und strukturelle Korruption. Die Grünen und die staatliche Parteienfinanzierung", in: Wewer, G. (Ed.): *Parteienfinanzierung und politischer Wettbewerb* (Opladen: Westdeutscher Verlag): 289–333.

Bachrach, P.; Baratz, M. S., 1970: *Power and Poverty* (Oxford: Oxford Univesity Press).

Panebianco, A., 1982: *Modelli di partito* (Bologna: Il Mulino).

Pareto, Vilfredo, 1916, 1964: *Trattato di sociologia generale*, Vol. 2 (Milan: Comunità).

Michels, Robert, ^{4}1989: *Soziologie des Parteiwesens* (Stuttgart: Kröner).

Salisbury, R. H. et al., 1987: "Who works with Whom: Interest Group Alliances and Opposition", in: *American Political Science Review*, 81: 1217ff.

Sartori, Giovanni, 1987: *The Theory of Democracy Revisited* (Oxford: Chatham House).

Therborn, G., 1980: *What Does the Ruling Class Do When It Rules?* (London: Verso).

Feist, U.; Liepelt, K, 1983: "Neue Eliten in alten Prteien", in: Kaase, Max; Klingemann, H.-D. (Eds.): *Wahlen und Politisches System* (Opladen: Westdeutscher Verlag).

Wassermann, Rudolf, 1986: *Die Zuschauerdemokratie* (Düsseldorf: Econ).

Welsh, William A., 1979: *Leaders and Elites* (New York: Holt, Rinehart & Winston).

Wessels, B., 1991: *Erosion des Wachstumsparadigmas* (Opladen: Westdeutscher Verlag).

About the Author

Klaus von Beyme (Germany) is Professor Emeritus of Political Science at the Faculty of Economic and Social Sciences of the University of Heidelberg. He studied Political Science, History, History of Art and Sociology at the Universities of Heidelberg, Bonn, Munich, Paris and Moscow (1956–1961). From 1961 to 1962, he was a Research Fellow at the Russian Research Centre of Harvard University and Assistant to Carl Joachim Friedrich. He obtained his doctorate (1963) and his habilitation (1967) from Heidelberg University. He was a Full Professor at the Tübingen University (1967–1973) and in 1971 was also its Rector a short time. From 1974 to 1999, Beyme was a Full Professor of Political Science at the University of Heidelberg.

He was President of the *German Society for Political Science* (DVPW, 1973–1975) and of the *International Political Science Association* (IPSA, 1982–1985). He taught as Visiting Profes-sor at *Stanford University* (California) 1979 and at the *École des Sciences Politiques* in Paris (1985) and at the *University of Melbourne*, in Australia (1989). He served a member of the Research Council at the *European University Institute* in Florence (1983–1990), of the Board of the Commission for Research into Social and Political Changes in the New Federal States (1990–1993) and he is member of the *Academia Europaea* and of the *Berlin-Brandenburg Academy of Sciences*. He received many awards for his scientific achievements, including Honorary Membership of the Humboldt University in Berlin (1995), the University Medal of the University of Heidelberg (1998), an Honorary Doctorate from the University of Bern (2001). He was honoured with the Schader Prize (2008). In 2010 he received a Honorary Professorship at Lomonosov University in Moscow. In 2012, at the 22nd World Congress of Political Science of the International Political Science Association (IPSA) in Madrid he was awarded

K. von Beyme, *Klaus von Beyme*, SpringerBriefs on Pioneers in Science and Practice 14,
DOI: 10.1007/978-3-319-01535-4, © The Author(s) 2014

the Prize of the Foundation Mattei Dogan awarded by the International Political Science Association for High Achievement in Political Science. In Germany 41 % of political scientists considered him as the most important representative of their discipline and he was ranked second regarding "professional political importance". Twice (1988, 1998) Klaus von Beyme was amongst the most important representatives of these fields of research: *Political Theory, Political Philosophy and History of Ideas* and for *Comparative Political Science, System Comparison.* His numerous books and translations in many languages are listed in chapter 2 and some book titles are at: http://www.afes-press-books.de/html/SpringerBriefs_PSP_vonBeyme.htm.

About the Book

Klaus von Beyme, a highly distinguished German political scientist is a "Pioneer in the Study of Political Theory and Comparative Politics". On the occasion of the receipt of the highly esteemed Mattei Dogan Award Session during the XXII World Congress of Political Science in Madrid on 12 July 2012, in his laudatio Rainer Eisfeld portrayed Klaus v. Beyme as "Global Scholar and Public Intellectual". On the occasion of Klaus v. Beyme's 80th birthday this book offers a selection of his major previously published and new texts focusing on "Empirical Political Theory", "The Evolution of Comparative Politics, Revival of Normative Political Theory in Empirical Research", "Theodor W. Adorno—Political Theory as Theory of Aesthetics", "Historical Forerunners of Policy Studies", "Political Institutions—Old and New", "Representative Democracy and the Populist Temptation", "Political Advisors to Politicians", and "The Concept of Political Class: A New Dimension of Research on Elites?"

K. von Beyme, *Klaus von Beyme*, SpringerBriefs on Pioneers in Science and Practice 14,
DOI: 10.1007/978-3-319-01535-4, © The Author(s) 2014

MIX
Papier aus verantwortungsvollen Quellen
Paper from responsible sources
FSC® C105338

If you have any concerns about our products,
you can contact us on
ProductSafety@springernature.com

In case Publisher is established outside the EU,
the EU authorized representative is:
Springer Nature Customer Service Center GmbH
Europaplatz 3, 69115 Heidelberg, Germany

Printed by Libri Plureos GmbH
in Hamburg, Germany